SCIENCE IN SECONDS

HAZEL MUIR

Quercus

CONTENTS

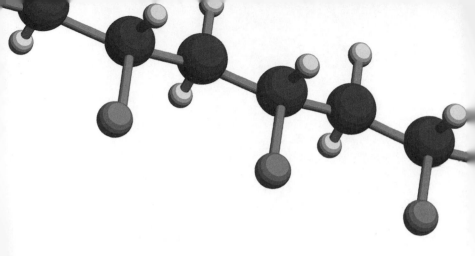

Introduction

Science is an amazingly powerful tool for understanding nature. For instance, it has revealed compelling evidence that the universe was born in a fiery explosion 13.7 billion years ago. Scientists have cracked the genetic codes of complex life and eradicated smallpox, a disease estimated to have killed up to 500 million people during the 20th century alone. When it comes to problem solving, scientific thinking is the best tool in the box.

But many science textbooks are tough going. *Science in Seconds* aims to be a readable, user-friendly introduction to key topics in science without getting technical. There's no reason why you can't pick up the gist of Einstein's relativity theories (see pages 16 and 18), for example, or how to clone a sheep (page 210) in a simple snack-sized introduction. Hopefully, this book will inspire even a few people to discover a subject they love and then dig deeper.

For simplicity, the book is structured into the traditional subjects such as physics, chemistry and biology, with an index for easy navigation. Narrowing down the topics to 200 wasn't easy, but the book includes must-have basics – for instance how biological cells divide and how lasers work – as well as some of the more cutting-edge fields such as stem cell therapy and the flourishing hunt for weird and wonderful alien planets beyond our own solar system.

These modern adventures remind us that science is not just about rote learning long-established theories – a scientist's real job is to find out things we don't already know. How can we prevent devastating diseases and climate change? Why does life exist at all? Most of the matter in the universe remains stubbornly unidentified, while more people have been to the Moon than to the deepest parts of our oceans. The real buzz of science lies in detective work, which will keep future scientists busy for a long time to come.

Motion

In physics, the motion of an object is described in terms of quantities such as velocity, acceleration and displacement, the distance of a moving object from its origin. Velocity is a vector quantity, specifying not only an object's speed but also its direction, while force is a measure of the push or pull that an object needs to change its velocity, resulting in acceleration – the rate of change of velocity over time.

Newton's laws of motion (see page 10) define the relationship between force and acceleration for everyday objects such as

cars and aircraft, which travel at speeds much slower than the speed of light. Momentum is defined as the product of an object's mass and its velocity. It is a 'conserved quantity', so that in the absence of any other influences, two snooker balls that bounce off each other will have the same total momentum before and after colliding.

An object's kinetic energy is equal to half its mass multiplied by the square of its speed. This quantity measures the work needed to accelerate the object to its given speed from rest.

1 Approaching ball of mass m and velocity v has momentum of $m \times v$

2 Second ball, also of mass m, is stationary, with zero momentum

3 As the balls collide, the first ball stops dead

4 All of the first ball's momentum is transferred to the second ball, which moves away with velocity v

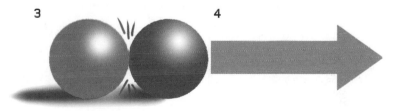

Newton's laws of motion

Isaac Newton's three laws of motion, first published in 1687, describe the relationship between a force acting on a body and the body's motion due to that force.

The first law says that a body moving with a given speed will maintain that speed in a straight line unless a force acts upon it – no force means no acceleration. The second law states that a force (F) will make a body accelerate by an amount (a) that's inversely proportional to its mass (m): $F=ma$. The third law says that whenever one body applies a force (the 'action' force) to a second one, the second one simultaneously applies an equal and opposite 'reaction' force on the first. Stepping off a boat onto a pier will make the boat move away, for instance.

Newton showed that these laws neatly explain the orbits of the planets around the Sun when combined with Newtonian gravity (see page 14). But they are not valid for objects moving at very high speeds or in very intense gravitational fields, when relativity theory is required (see pages 16 and 18).

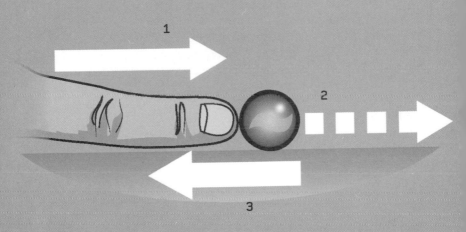

1 Finger applies force *F* to marble
2 Marble accelerates according to *F=m x a*
3 Finger experiences a reaction force *F* in terms of pressure

Centripetal and centrifugal forces

A centripetal force is one that makes a body move in a curved path. Gravity is an example of a centripetal force in Newtonian gravity (see page 14), making a planet orbit a star by continually accelerating the planet towards the star at the orbit's centre. Without this centripetal force, the planet would fly off into space in a straight line.

When you whirl a tennis ball over your head on a string, the ball feels a centripetal 'pull' force. The centripetal force is often confused with the centrifugal (outward) force, which can be a 'fictitious' force. It accounts for the sense of being pushed outwards when looping-the-loop on a roller coaster.

The centrifugal force can also be a reaction force to a centripetal force, according to Newton's third law of motion (see page 10). In the case of a tennis ball on a string, the whirling ball exerts an outward centrifugal force on the person spinning it.

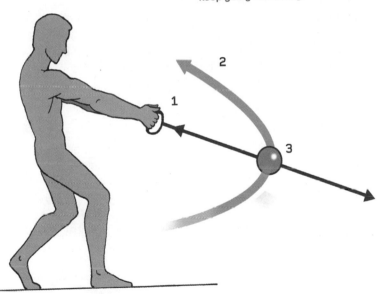

1 Inward force of tension between ball and athlete

2 Curved path of ball around athlete

3 Outward 'centrifugal' force is caused by the ball's tendency to keep going in a straight line

Newtonian gravity

Isaac Newton's law of universal gravitation, published in 1687, was the first clear mathematical description of how bodies such as planets and stars attract each other under their mutual gravitational pull.

Newton's inspiration for the theory came from watching an apple falling from a tree. A falling apple accelerates towards the ground, so Newton reasoned from his laws of motion (see page 10) that there must be a force, which he called gravity, acting on the apple. This force might have a huge range and could also be responsible for the orbit of the Moon around the Earth, if the Moon had just the right speed to remain in orbit despite constantly 'falling' towards the Earth.

He went on to show that the gravitational force between two massive objects is directly proportional to the product of their masses and weakens with the square of the distance between them. But troublingly, the theory didn't explain why the force was transmitted across empty space. This problem is resolved in Einstein's general relativity theory (see page 18).

1 Acceleration due to gravity
  = 9.8 m/s (32.2 ft/s) per second
2 Force of gravity on a 1 kg (2.2 lb)
  object = 9.8 newtons

1

2

Special relativity

Special relativity is the theory of motion published by Albert Einstein in 1905. Einstein developed it from two basic principles: the laws of physics must be the same for any observer moving at a constant velocity, and the speed of light is always the same, regardless of the speed of the light source.

Relativity abandons the idea that it's possible to have a universal standard of time and space. Instead, the length of an object or time interval depends on who is measuring it. Take the case of a train moving at close to the speed of light relative to an observer. The observer would perceive the train to be shorter than the passengers on board would measure, while the observer would see a clock on the train run slow.

This is not just an illusion – measurements show that unstable particles moving fast through Earth's atmosphere decay much more slowly than they do at rest in a laboratory. Special relativity forbids massive objects from travelling as fast as the speed of light in a vacuum, which would require an infinite amount of energy.

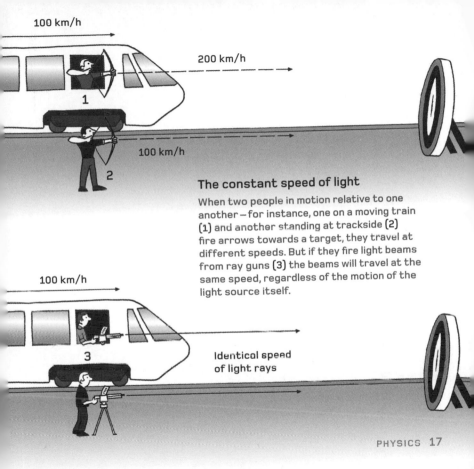

100 km/h

200 km/h

100 km/h

The constant speed of light

When two people in motion relative to one another – for instance, one on a moving train (1) and another standing at trackside (2) fire arrows towards a target, they travel at different speeds. But if they fire light beams from ray guns (3) the beams will travel at the same speed, regardless of the motion of the light source itself.

100 km/h

Identical speed of light rays

General relativity

General relativity is Einstein's theory of gravity, developed by 1915. Unlike Newtonian gravity (see page 14), Einstein's theory views gravity as a natural upshot of the geometry of curved space, and ditches the notion that gravity is 'action at a distance'. Large masses like planets move in response to the curvature of space–time, distorted by mass itself. Matter tells space how to curve; curved space tells matter how to move.

It's difficult to visualize in three dimensions, but it helps to imagine a star's mass making a depression in a two-dimensional sheet. A nearby planet would be forced to curve around it like a ball in a roulette wheel.

Some predictions of general relativity are different from those of Newtonian gravity. Although both theories predict that the Sun's gravity bends light from background stars, which can be measured when sunlight is blocked during a solar eclipse, Einstein's theory predicts twice as much deflection as Newton's. Measurements show general relativity is correct on this, and it has passed all other tests so far with flying colours.

1 True position of star
2 Apparent position of star
3 Distorted space-time around the
  Sun deflects path of starlight
4 Position of observer on Earth

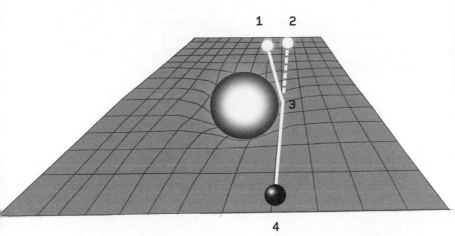

Temperature and pressure

Temperature is a measure of how hot an object is, which reflects the amount of kinetic energy in its molecules. For common purposes, most countries measure temperatures using the Celsius scale of temperature, which sets water's freezing point at zero degrees (0°C) and its boiling point at 100°C. The US uses the Fahrenheit scale, in which water freezes at 32°F and boils at 212°F.

Matter can be cooled by reducing the kinetic energy of its molecules, but the laws of thermodynamics (see page 28) predict that there is a minimum possible temperature, which turns out to be −273.15°C (−459.67°F). This is 'absolute zero', where particles would theoretically be motionless.

Pressure is the force exerted by one substance on another substance, per unit area. The pressure of a gas is the force the gas exerts on the walls of its container. The standard unit of pressure is the pascal (1 newton of force per m^2). The typical air pressure at sea level on Earth is about 100,000 pascals.

1 Random motions of gas molecules in container

2 Speed of molecules increases with temperature

3 Collisions of gas molecules with container walls and other molecules exert pressure

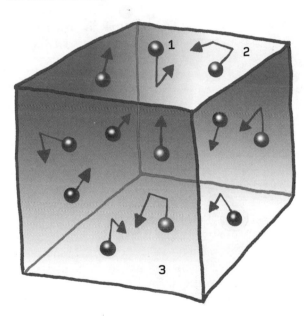

Heat transfer

Heat can be transmitted through matter in three ways: conduction, convection and electromagnetic radiation (see page 52). Unlike conduction and convection, radiation can transmit energy across empty space.

Conduction is the mechanical transfer of heat through matter from a warm part to a cooler one without any bulk motion. In gases and liquids, heat conduction occurs due to collisions and diffusion of molecules during their random motion. In solids, molecules conduct heat by vibrating against each other or when free electrons carry kinetic energy from one atom to another. Metals are the best conductors of heat.

Liquids and gases also transfer heat in convection currents, which involve bulk fluid motion. For instance, a hot bubble of gas in the Sun's atmosphere can carry heat into a higher, cooler layer before cooling and sinking. Heat transfer can also occur when radiation carries energy between one object and another. For instance, sunlight heats the Earth by making molecules in the Earth's atmosphere and surface vibrate.

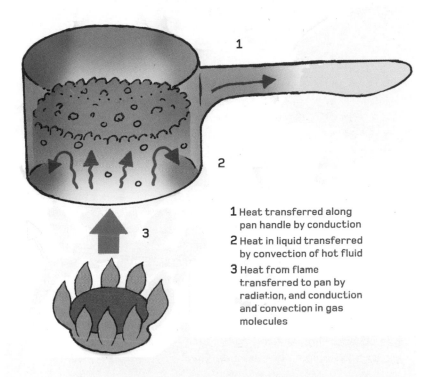

1 Heat transferred along
  pan handle by conduction
2 Heat in liquid transferred
  by convection of hot fluid
3 Heat from flame
  transferred to pan by
  radiation, and conduction
  and convection in gas
  molecules

Brownian motion

Brownian motion describes the jittery, random motion of relatively large particles suspended in a fluid or gas, such as smoke particles in air. It is named after Robert Brown, a Scottish doctor and botanist who studied it in detail in 1827.

Brown noticed that pollen grains in water jiggled about, following zigzag paths. Later in 1905, Albert Einstein showed that this Brownian motion can be predicted mathematically by assuming these large, suspended particles are constantly bumped by smaller fluid molecules moving due to their own thermal energy. One prediction was that the displacement of a suspended particle from its origin over time should be proportional to the square root of the time elapsed.

Experiments by French physicist Jean Perrin soon confirmed that Einstein's predictions were correct, indirectly proving that molecules and atoms exist, despite the fact that they were too small to be seen directly. This might seem obvious now, but it was still common at the time to believe that matter was not grainy and could be divided indefinitely.

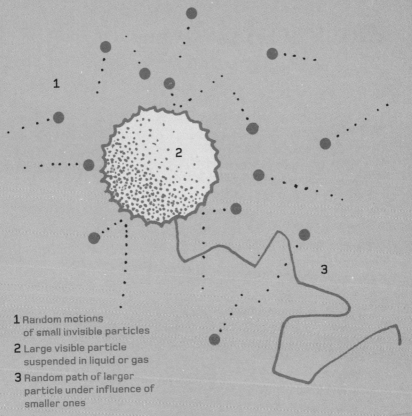

1 Random motions
  of small invisible particles

2 Large visible particle
  suspended in liquid or gas

3 Random path of larger
  particle under influence of
  smaller ones

Work and energy

Work refers to an activity involving a force and movement, while energy is the capacity for doing work – a bit like a 'currency' that gets used up in the process. In the context of a moving object, the work done by a force equals the force multiplied by the distance moved.

In the context of thermodynamics, work has a more complex definition. It refers to energy transferred to a gas, for instance, but only if that energy causes a macroscopic change to the gas, perhaps making it expand its volume against an external pressure. It doesn't include the input of heat energy if the heat merely increases the microscopic thermal motions of particles.

The work done to compress a gas in a container with a movable piston is approximately equal to the gas pressure times the volume change. The change in internal energy of a gas is equal to the heat added minus the work done by the gas, which is one way of stating the first law of thermodynamics (see page 28).

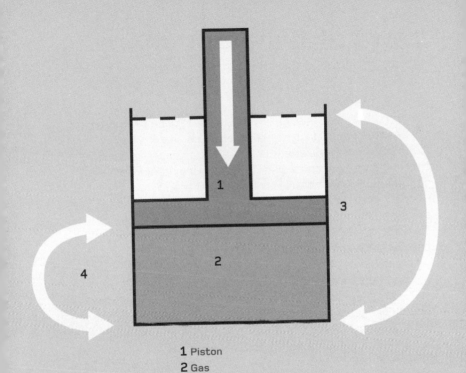

1 Piston
2 Gas
3 Volume before compression
4 Volume after compression

Laws of thermodynamics

The four laws of thermodynamics define the relationships between quantities like temperature and work in 'thermodynamic systems' – a loose term for any matter with thermal energy, such as gas molecules in a container.

'Thermal equilibrium' describes the state of two systems in contact with each other, which have no net exchange of energy because they've reached the same temperature. The 'zeroth law' of thermodynamics says that two systems in thermal equilibrium with a third one must also be in thermal equilibrium with each other. Scientists felt the need to state the intuitively obvious zeroth law after they adopted the other three.

The first law says energy in an isolated system is conserved. Chemical energy might change into kinetic energy, but the total stays the same. The second law says that because energy varies in its quality or ability to do useful work, the entropy of an isolated system – a measure of the energy input that doesn't do mechanical work – always increases. The third law says minimum entropy occurs at absolute zero (see page 20).

According to the zeroth law of thermodynamics, if objects (1) and (3) are in thermal equilibrium and objects (1) and (2) are also in equilibrium, then (2) and (3) must also be in equilibrium with each other

Phases of matter

Classically, matter can exist in three phases: solid, liquid and gas. Traditionally, solids are defined as matter with a fixed volume and shape, containing closely packed particles. Liquids keep the same volume but flow to fill the bottom of a container, while gases expand to occupy all available volume.

Phase transitions can occur due to alterations in pressure or temperature. At normal atmospheric pressure, pure water melts from solid ice into a liquid above 0°C (32°F) and boils into water vapour at 100°C (212°F). The energies of individual water molecules in a boiling kettle are not identical but follow a bell curve, which means that liquid and gas phases can co-exist. At the so-called triple point of a substance, all three phases can co-exist. For instance, water ice, liquid and vapour can mingle in a container at 0.01°C (32.02°F) at very low pressures.

Plasma, a searingly hot ionized (electrically charged) gas, is often called a fourth state of matter. It streams out from stars like the Sun into interstellar space. More exotic states of matter include Bose—Einstein condensates (see page 80).

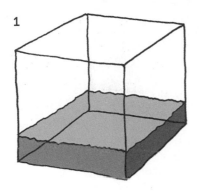

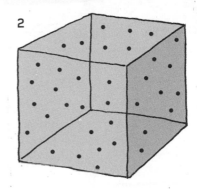

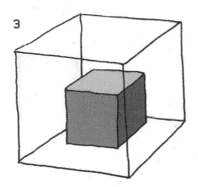

1 Liquid maintains same
volume and flows to the
bottom of container

2 Gas fills all available
space in container

3 Solid maintains both
the same shape and
volume regardless of the
shape of container

Surface tension

Surface tension results from the inward pull of molecules on the surface of a liquid, making them adopt the smallest surface area possible. It effectively makes the surface stronger, allowing a small object such as a sewing needle to effectively 'float' on water, even though the object may be much denser than the liquid.

In the bulk of a liquid, molecules face a tug of war in which they're pulled equally in all directions by neighbouring molecules, so the forces on them cancel out. But surface molecules lack upward force, so they're pulled together and down, making the surface contract to its minimum size.

Surface tension holds water droplets together and would make them spherical in the absence of other forces such as gravity, because a sphere has the smallest surface to volume ratio. Many animals take advantage of surface tension on ponds. Common insects called water striders, or pond skaters, rely on it to walk on water and sense vibrations from nearby prey using sensitive hairs on their legs and bodies.

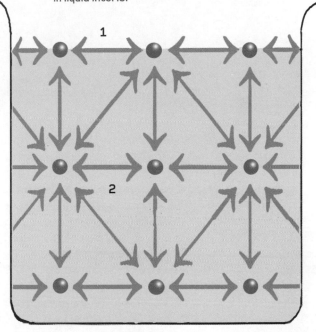

1 Attractive forces pull surface particles inwards and create tension across surface

2 Equal attractive forces between molecules in liquid interior

Archimedes' principle

Archimedes' principle states that the buoyant force on an object submerged in a fluid (liquid or gas) is equal to the weight of the fluid that the object has displaced. It implies that an object will sink in a fluid if its average density is greater than that of the fluid.

Archimedes was a Greek scientist and engineer who lived during the 3rd century BC. Later historians suggest that he was tasked with determining whether a crown supposedly crafted from pure gold also contained some cheaper silver. While taking a bath, Archimedes noticed that the water level rose when he got in, and realized that by placing the crown in water and measuring the displaced water volume, he could establish the volume of the crown, and thus calculate its density and purity without damaging it.

Legend has it that Archimedes then ran down the street naked shouting 'Eureka!', Greek for 'I have found it'. His principle explains why ships float and why hot-air balloons rise – warm air within a balloon is less dense than the cooler air outside.

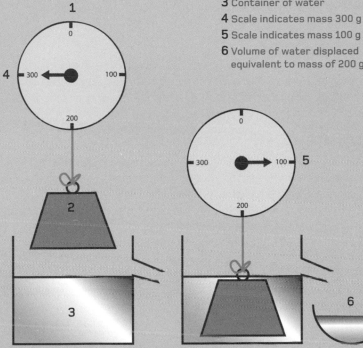

1 Weighing scale
2 Mass
3 Container of water
4 Scale indicates mass 300 g
5 Scale indicates mass 100 g
6 Volume of water displaced
  equivalent to mass of 200 g

Fluid dynamics

Fluid dynamics is the science of how fluids (both liquids and gases) flow. It's essential for many practical applications, including the design of efficient aircraft, ships and oil pipelines as well as weather forecasting.

A boat moving through water encounters two main kinds of resistance – inertial forces from the water (effectively the water's resistance to motion) and viscosity or stickiness. In fluid dynamics, the 'Reynolds number' expresses the relative importance of these factors in flows across a surface, such as a ship's hull or a pipeline. A low Reynolds number gives smooth fluid motion, while turbulent flow with chaotic eddies and vortices occurs at high Reynolds numbers.

One key concept in fluid dynamics is the Bernoulli effect: the faster a fluid flows, the lower its pressure. The curved upper surfaces of aircraft wings are shaped to force air to follow a longer path over the top of the wing, speeding it up. This lowers pressure above the wing and creates a net upward force, or lift.

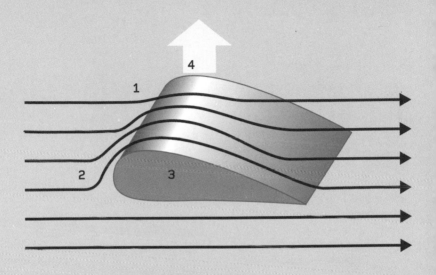

Fluid dynamics of aircraft

1 Faster flow of air diverted over top of wing

2 Slower flow of air passing below wing

3 Area of higher pressure

4 Area of lower pressure creates lift

Wave types

A wave is a disturbance that propagates through empty space or a medium such as air or water, usually transporting energy as it travels.

In 'transverse waves', the disturbance is at right angles to the wave's direction of motion. Electromagnetic radiation, including visible light, is a form of transverse wave in which magnetic and electric fields oscillate at right angles to the wave's direction of travel. In 'longitudinal waves', the disturbance is parallel to the wave's direction. These include sound waves in gases and liquids. Water waves are an example

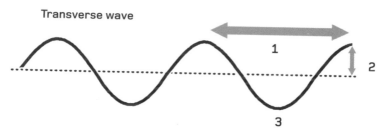

Transverse wave

of a wave that is both transverse and longitudinal – a floating cork will move in a circle as a wave washes past it.

Waves are characterized by their wavelength (the distance between peaks or compressions), frequency (the rate at which waves pass a given point and amplitude or intensity. Standing or stationary waves occur when waves are held in a fixed position – for instance, when a guitar string vibrates. Such waves always involve a whole or half-number of waves, and hence the length of the string determines the wavelengths it can maintain.

1 Wavelength: distance between successive wave peaks or troughs

2 Amplitude: height of waves or magnitude of disturbance

3 Frequency: number of peaks or troughs passing a fixed point in a second

Longitudinal wave

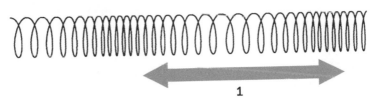

Sound waves

Sound waves are pressure oscillations that propagate through a gas, liquid or solid – sound can't travel in empty space. In gases and liquids, sound is a longitudinal wave (see page 38), but transverse sound waves can pass through solids.

People hear sound because it makes our ear drums vibrate. These vibrations pass through the inner ear to nerve cells that send signals to the brain, which interprets them as sound. A higher wave frequency means that the air pressure fluctuation switches back and forth more quickly, and we hear this as a higher pitch. Human hearing is normally limited to frequencies between 20 and 20,000 hertz (repeating wave cycles per second) and the upper limit tends to drop with age.

The speed of sound depends only on the transmitting medium. In air with a temperature of 20°C (68°F) at sea level, sound travels at about 343 m/s (767 mph). The intensity of sound is measured in decibels, with a typical conversation registering about 60 decibels and motorbike engines topping 100 decibels.

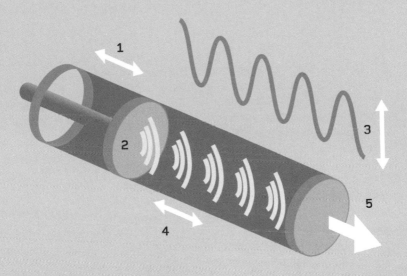

Sound waves in a piston

1 Vibrations generate sound
2 Piston vibrates
3 Intensity: peak-to-peak pressure change

4 Wavelength: distance between peaks
5 Direction of wave propagation

Doppler effect

The Doppler effect describes changes in the frequency of waves depending on how the wave source is moving relative to an observer. It explains why the siren of a fire engine sounds higher in pitch as it comes towards us, then sounds lower after the vehicle has driven past.

When a source of sound waves moves towards an observer, each successive wave emitted comes from a position closer to the observer, who hears it more quickly due to the shorter travel time. Effectively, the waves bunch together creating an increase in frequency. Conversely, when the wave source moves away, successive waves are emitted from greater distance. The waves stretch out, making their frequency drop.

The effect is named after Austrian physicist Christian Doppler, who described the effect for light waves in 1842. Frequency also determines the colour of light, so the Doppler effect alters the colour of a light source approaching an observer or receding at very high speed – a green light appears more blue when approaching and more red when receding.

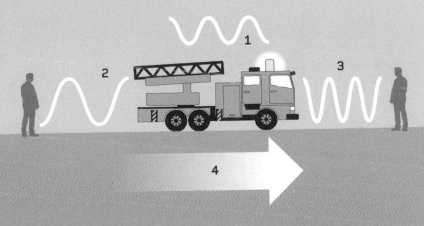

1 Sound heard by crew on fire engine

2 Lower-frequency waves reach observer behind

1

2

3

3 Higher-frequency waves reach observer in front

4 Direction of motion

4

Electric charge

Electric charge is a property of many standard model particles (see page 84), including the electron. This property makes them feel a force from other charged particles. Electrical charge can be either negative or positive, with negatively charged particles attracting positively charged ones while repelling their own kind.

The unit of electric charge is called the coulomb (C); 1 coulomb is the charge transported per second by an electric current of 1 ampere. The negative charge of an electron is -1.602×10^{-19}C. For simplicity, the electron charge is often denoted as -1, while that on a positively charged proton is $+1$.

Electric charge plays a pivotal role in our very existence, allowing solid structures like the Earth, buildings and animals to exist. Atoms are mostly empty space, but they don't fall through each other due to repulsion between electrons in neighbouring atoms. Charged particles zinging around in the Sun's atmosphere also play a crucial role by generating the radiation that keeps our planet's surface warm and hospitable.

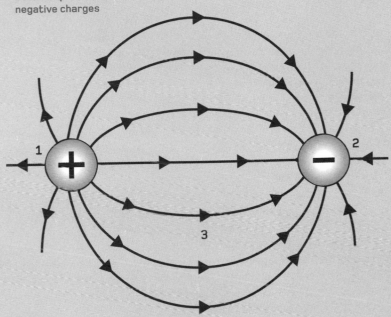

1 Positive charge
2 Negative charge
3 Electric field created
   between positive and
   negative charges

Electric current

Electric current is the flow of electric charge, carried by moving electrons. A current flows through a conducting material such as copper wire when it is connected to the positive and negative terminals of a battery, which applies an electrical potential difference or voltage. Electrons in the wire then drift towards the positive terminal.

National electricity grids deliver alternating electric current that reverses periodically, usually 50 or 60 times a second. The unit of electric current is the ampere; 1 amp is the flow of 1 coulomb of charge per second.

Electrical resistance is the extent to which a material resists the flow of electric current, and is measured in ohms. Metals such as silver and copper have low resistance, allowing current to flow easily, while plastics and wood have high resistance, making them poor conductors. The current in a wire is equal to the voltage applied across it divided by the resistance, while electrical power – the energy transmitted per unit time – is the product of the voltage and current.

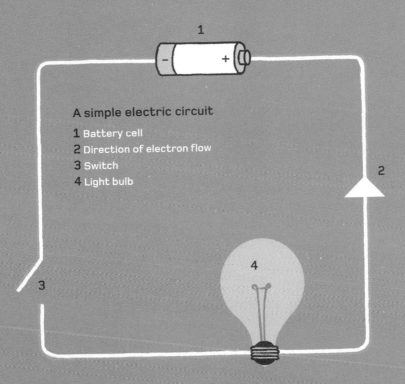

A simple electric circuit

1 Battery cell
2 Direction of electron flow
3 Switch
4 Light bulb

Magnetism

Magnetism is a property of materials that makes them experience a force in a magnetic field. It explains why iron filings line up in ordered patterns near a bar magnet for instance, and why fridge magnets stick to fridges.

Bar magnets are strips of magnetized metal, usually iron, that form a 'dipole' field with a north and south pole. Opposite poles attract each other, while like poles repel. The magnetic field of a permanent magnet arises because electrons inside it generate their own tiny magnetic fields due to an intrinsic property called spin, and in materials like iron, the spins of unpaired electrons tend to line up.

Scientists have known since the early 1800s that there is a deep connection between magnetism and electric current. For instance, an electric current flowing through a coil of wire creates a dipole magnetic field similar to that of a bar magnet. Modern electromagnets have achieved record-breaking magnetic fields of 35 teslas, where 1 tesla is about 20,000 times stronger than the Earth's magnetic field (see page 258).

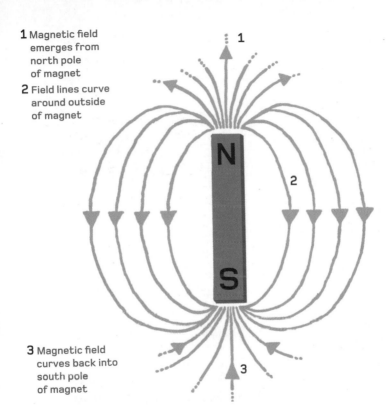

1 Magnetic field emerges from north pole of magnet

2 Field lines curve around outside of magnet

3 Magnetic field curves back into south pole of magnet

Induction and capacitance

Induction occurs when a conducting material moves through a magnetic field. In 1831, the English scientist Michael Faraday showed that this makes an electric current flow through the conductor. Induction underpins the operation of electrical equipment from electric motors to electricity generators.

Dynamo generators convert rotational motion produced by a turbine, for instance, into electricity, while electric motors do the opposite, creating rotational motion from an electric current. In both cases, the motion, magnetic field and electric current are all perpendicular to each other in directions illustrated by left-hand and right-hand rules, memory aids invented by British engineer John Ambrose Fleming.

Electrical circuits also have 'self-inductance' — changes in current in a wire will generate a changing magnetic field, which in turn induces a current. Inductors are electrical components designed to store energy in induced magnetic fields, while 'capacitors' store energy in electric fields. In simple capacitors, opposite electric charge builds up on two parallel plates.

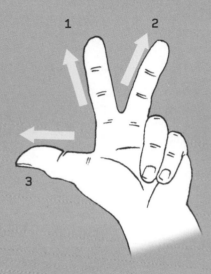

Fleming's left-hand rule for motors

1 Forefinger = field
2 Centre finger = current
3 Thumb = motion

Fleming's right-hand rule for generators

4 Thumb = motion
5 Forefinger = field
6 Centre finger = current

Electromagnetic radiation

Electromagnetic radiation is a form of energy that can travel through empty space and includes visible light. It also includes gamma rays, which can cause radiation sickness by damaging cells, and radio waves, which are vital for wireless communication technologies.

Electromagnetic radiation a transverse wave (see page 38) consisting of oscillating electric and magnetic fields. In a vacuum, the waves always travel at 300,000 km/s (671 million mph),

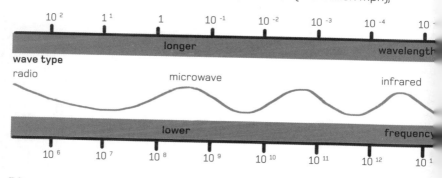

10^2	1^1	1	10^{-1}	10^{-2}	10^{-3}	10^{-4}	10^-

longer   wavelength

wave type

radio   microwave   infrared

lower   frequency

10^6	10^7	10^8	10^9	10^{10}	10^{11}	10^{12}	10^1

but their wavelengths vary enormously. Gamma rays have tiny wavelengths, often smaller than an atom, while radio waves can be thousands of kilometres across.

We only see a small part of the electromagnetic spectrum – visible light spanning the rainbow of colours from violet to red. Visible sunlight passes through the Earth's atmosphere and reflects off objects allowing us to see them. Many insects, fish and birds can also see ultraviolet radiation, which attracts bees to flowers. Gamma rays are much more penetrating and can pass through several centimetres of lead.

The electromagnetic spectrum stretches from low-frequency radio waves to high-frequency gamma rays.

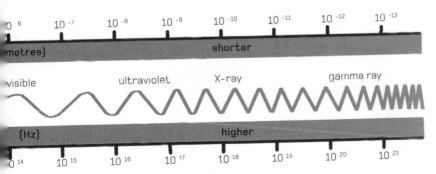

Photons

The photon is the quantum of electromagnetic radiation (see page 52) and the basic 'unit' of light. In a vacuum, photons all move at the same speed: 300,000 km/s (671 million mph).

Light seems to have a split personality – it can be thought of both as a wave and a stream of particles (see page 68). Albert Einstein nailed the case for the particle-like nature of light when he explained the 'photoelectric effect', in which light shining onto a piece of metal makes the metal eject electrons. Strangely, a dim blue light has this effect, but a red light, no matter how bright, does not. Einstein realized that this is because light consists of discrete energy packets. An individual photon of blue light has enough energy to dislodge a single electron from a metal, but red photons do not, regardless of how many there are.

Photons have no mass or electric charge, but they carry momentum. The energy of a photon of light is proportional to the light's frequency, with gamma-ray photons carrying billions of times more energy than radio photons.

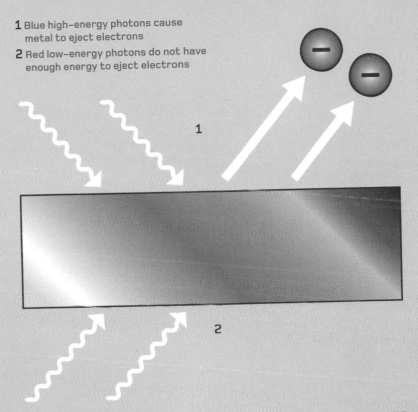

1 Blue high-energy photons cause metal to eject electrons
2 Red low-energy photons do not have enough energy to eject electrons

1

2

Lasers

Laser light differs from ordinary light because it is much more organized – a bit like a marching army compared to a bustling crowd. While light from a light bulb contains many wavelengths, laser light contains only one. It forms a much narrower beam and is 'coherent', meaning the waves all travel in lockstep, with their crests and troughs lined up.

Lasers (short for Light Amplification by Simulated Emission of Radiation) were first developed in the 1960s. Laser light is produced when atoms or molecules are excited to higher energy levels in a cavity, then zapped with photons of a specific energy that the hyped-up particles can emit to relax back to their normal state. This makes the particles emit clones of these photons, with exactly the same properties, in a chain reaction to create a laser beam.

Lasers have a host of everyday applications, such as reading information in DVD players and barcode scanners, and as a tool for hospital surgery. Future gamma-ray lasers could focus a million times more energy than the current generation.

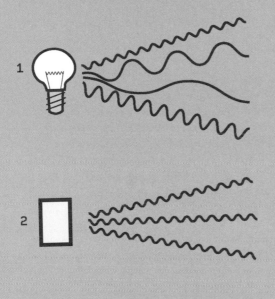

1 Light from a normal source is a mix of wavelengths and frequencies diverging from source

2 Light rays from a onochromatic light source have identical wavelengths but are not in lockstep, and diverge from source

3 Light rays from a coherent laser light source are monochromatic and tightly aligned, with waves in lockstep

Reflection and refraction

In many situations, light can be thought of simply as a transverse wave (see page 38) that travels in a straight line until it encounters an obstacle. Light reflects in a simple way off smooth surfaces like mirrors, obeying a law that the angle of reflection (measured from the 'normal', a line perpendicular to the mirror) equals the angle of incidence or approach.

Refraction is the process that happens when light travels from one transparent medium, such as empty space or air, into another one, such as water. Light travels slower in water than in a vacuum, and this change of speed is responsible for the bending of light, or refraction. The ratio of the vacuum speed to the water speed is called the 'refractive index' of water, which is about 1.33.

Light bends towards the normal when passing into a denser medium and away from the normal when entering a less dense one. Glass lenses in spectacles and optical instruments like telescopes are specially shaped to refract light rays onto the required paths for correcting poor vision or focusing starlight.

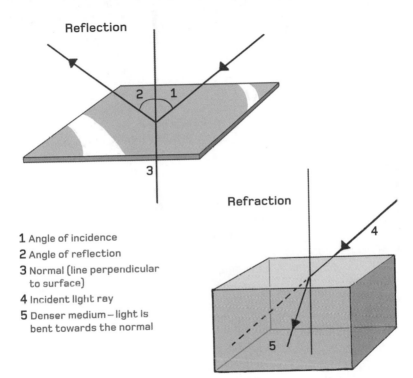

Reflection

1 Angle of incidence
2 Angle of reflection
3 Normal (line perpendicular to surface)
4 Incident light ray
5 Denser medium — light is bent towards the normal

Refraction

Diffraction

Diffraction describes the way waves bend around obstacles they encounter. A classic example is the way that water waves fan out when long, straight wavefronts encounter a narrow opening, then emerge as small circular waves beyond it.

The effect is easy to demonstrate by creating ripples in a tray of water containing a barrier with two gaps in it. Straight waves approaching the barrier create small circular waves beyond the two gaps, and as they move outwards they undergo interference (see page 64), with crests amplifying each other while crests and troughs cancel out. Light diffraction patterns can be unintuitive and are best seen with laser light. The diffraction pattern of a square hole is cross-shaped, while a circular aperture produces a series of concentric circles.

Diffraction by water droplets or ice crystals in thin clouds sometimes creates a beautiful bright ring around the Sun or the Moon. But diffraction is largely a nuisance to optical instrument designers, setting fundamental limits on the clarity of images taken by cameras, microscopes and telescopes.

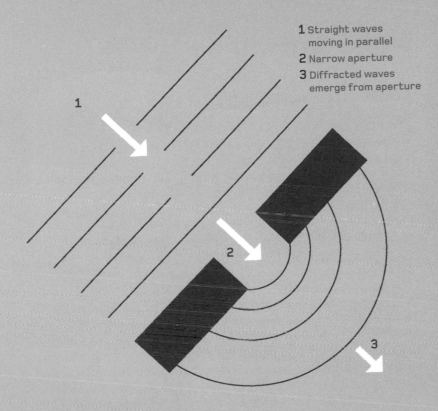

1 Straight waves moving in parallel
2 Narrow aperture
3 Diffracted waves emerge from aperture

Polarization

Polarization is a property of transverse waves (see page 38) that are restricted in their directions of oscillation. It's most frequently discussed in the context of electromagnetic radiation, including normal light, which can be polarized using a filter that only transmits light beams 'waving' in one plane.

Light consists of electric and magnetic fields oscillating perpendicular to each other, but the electric fields in ordinary light from the Sun or a torch oscillate in all possible planes. A linearly polarized light beam is one in which the electric field oscillations are restricted to one plane. It's also possible to create circularly polarized light, in which the electric field oscillations continually rotate like the thread of a corkscrew as the light beam moves through space.

Light reflected from surfaces such as a flat road or smooth water tends to be horizontally polarized. Polarized sunglasses reduce reflected glare using filters that contain long-chain molecules. These preferentially absorb horizontally polarized light, so that only the vertical component passes through.

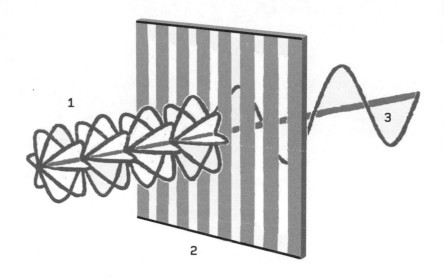

1 Normal unpolarized light oscillates in multiple planes
2 Polaroid filter acts as a narrow grille
3 Emerging polarized light oscillates in just one plane

Interference

Interference occurs when waves overlap. If you drop two stones in a puddle and watch the ripples spread out, you'll see them combine to form a distinctive pattern in which coinciding wave crests have amplified each other in 'constructive interference', as have matching troughs, while peaks and troughs cancel each other out in 'destructive interference'.

A thin film of oil on water can create colourful light interference patterns when sunlight reflects from the top of the oil as well as the oil–water boundary. The two reflections have followed different path lengths, so when they recombine, they interfere in a constructive or destructive way depending on the light's particular wavelength, or colour. This makes white light fan out into a rainbow of colours that changes with viewing angle. Light reflecting off the many grooves of a shiny CD or DVD creates colourful interference in a similar way.

Sound interference is noticeable when two tones have an almost identical pitch. This creates a warbling effect called beating due to constructive and destructive interference.

The dual-slit experiment in light waves

1 Monochromatic light source produces light of a single wavelength

2 Barrier with dual slits

3 Constructive interference where peaks meet peaks and troughs meet troughs

4 Destructive interference where peaks meet troughs

5 Interference pattern formed on screen

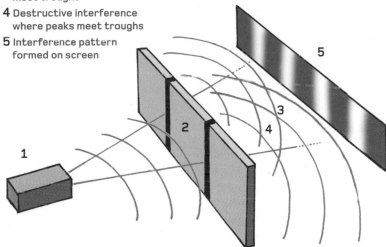

Quantum mechanics

Quantum mechanics is a branch of physics that describes the strange behaviour of matter and energy on the tiniest scales. Scientists developed it in the early 20th century when experiments revealed cracks in classical physics. It was clear that electrons orbited an atom's nucleus, for instance, but if they did so like planets orbiting the Sun, they should fall into the nucleus in a split second – which obviously doesn't happen.

Quantum mechanics invokes ideas like the Heisenberg uncertainty principle (see page 70) to explain the anomalies of tiny realms. One key concept is that properties of particles, such as the energies of electrons in an atom, can only change by discrete amounts – they are said to be 'quantized'.

The quantum world is strangely unpredictable. Everyday experience suggests you could take an electron, apply a force to it, then predict where the electron will be a second later. Quantum mechanics says this is impossible. You can estimate the likelihood of the electron reaching a given place, but until you measure its position, it's in all possible places at once.

In the 'Bohr model' of an atom, electrons orbit the atomic nucleus in quantized 'orbitals' – they can only move between orbitals by absorbing or emitting energy.

1 Electron absorbs energy and moves to a higher orbital

2 Electron emits energy in order to fall to a lower orbital

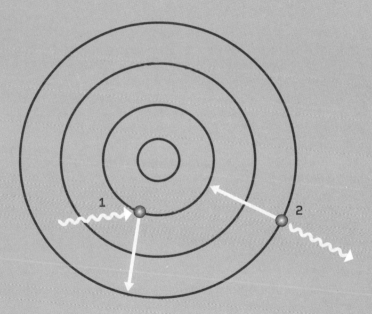

Wave-particle duality

Wave–particle duality describes the way matter and energy on the smallest scales can be viewed as both particles and waves. In normal life, we expect moving particles to behave like small projectiles, while waves spread out like ripples on a pond. In quantum mechanics, this distinction is blurred.

Electrons display this duality in the 'dual-slit experiment'. When electrons emerging from a source shine through two slits onto a phosphorescent screen, bright and dark bands form due to interference analogous to that seen in light waves (see page 64). What's more, even if the source is calibrated to produce just one electron at a time (which according to classical physics should only pass through one slit or the other, and hence only hit one of two areas of the screen) the interference pattern still builds up over long periods of time.

Bizarrely, the interference vanishes if the experiment is modified to detect which slit each electron is passing through – it's impossible to observe particle-like positional information and wave-like interference patterns simultaneously.

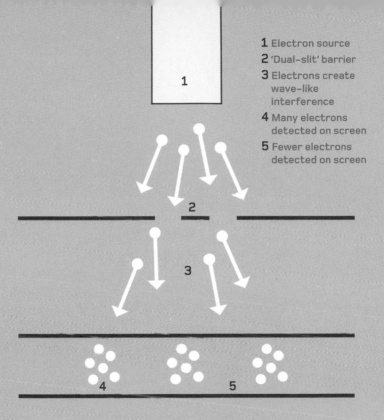

1 Electron source
2 'Dual-slit' barrier
3 Electrons create wave-like interference
4 Many electrons detected on screen
5 Fewer electrons detected on screen

Uncertainty principle

The Heisenberg uncertainty principle underlines the fuzziness of the quantum world. It states that certain pairs of properties, such as the position and momentum of a particle, can't both be determined with exact precision at the same time. The more precisely you know the particle's position, the less precisely you can know its momentum.

In 1927, the German physicist Werner Heisenberg published the principle, which stems from a particle's wave-like nature.

1 Uncertainty of position: the more accurately the object's wavelength is known the less accurately its position can be determined

The only wave with a definite position is concentrated at a single point, but such a wave has an indefinite wavelength, meaning indefinite momentum. Conversely, the only wave with a precise wavelength is infinitely long and has no definite position. So there are no states that simultaneously describe a particle's exact position and momentum.

Heisenberg's uncertainty principle quantifies this vagueness – the product of the uncertainty in position and momentum must be greater than or equal to 'Planck's constant' h (a tiny number equal to 6.6×10^{-34} joule seconds) divided by 4π.

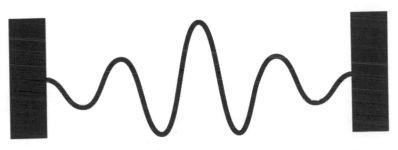

2 Uncertainty of wavelength: the more tightly constrained the object's location, the harder it is to pinpoint its wavelength

Schrödinger's cat

Schrödinger's cat is a thought experiment proposed by an Austrian physicist, Erwin Schrödinger, in 1935. He wanted to highlight a problematic paradox in quantum mechanics, that the properties of particles can't be determined until they're observed. The position of an electron, for instance, is a superposition of all possible positions until it is measured.

The thought experiment questions what would happen to a cat shut in a box with a device containing a radioactive nucleus and a lethal poison. If the nucleus decayed by emitting a particle, that would trigger the poison's release, killing the unfortunate moggie. But quantum theory forbids predictions of when the nucleus will decay. So is the cat is both dead and alive until we 'measure' its state by opening the box to look at it?

To this day, scientists debate various solutions to this paradox. Perhaps the simplest view is that quantum theory doesn't really create this paradox at all, because it clearly states that the only possible measurements are sensible ones, in this case a dead or alive cat, nothing in between.

A radioactive source (1) triggers the release of a deadly poison (2) on the decay of its nucleus. If the nucleus decays then the cat dies (3), but if the nucleus does not decay, the cat lives (4). The superposition of both states given by quantum mechanics suggests that until the system is observed, the cat can be both 'dead' and 'not dead'.

Quantum entanglement

Quantum entanglement is a weird effect in which two particles can be set up to 'know' what the other one is doing, even when they are separated by thousands of miles and have no way of communicating with each other.

The effect arises in quantum mechanics because it's possible to link the properties of two particles so that they will always be related. For instance, two light photons can be primed so

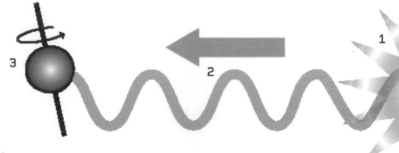

1 Pair of entangled particles are created in a laboratory
2 Particles are separated by a huge distance

that their polarization states are unknown, but when measured will be opposite. The two photons can zoom off into space in opposite directions with undefined polarizations, but when someone later measures the polarization of one photon, the second one will take on the opposite polarization. It's like instantaneous communication, faster than the speed of light.

Albert Einstein was suspicious of theories that allowed quantum entanglement, calling it 'spooky action at a distance'. But experiments prove that it really happens. Scientists have successfully transmitted entangled photons between sites on Spain's Canary Islands 140 km (87 miles) apart.

3 Particles remain entangled – when quantum information about one is measured...

4 ... the second particle will instantly 'collapse' into the complementary state

Casimir effect

In quantum mechanics, the Casimir effect describes the tiny attractive force acting between two parallel, uncharged conducting plates in a vacuum. It arises because a vacuum is not just empty space – it is seething with energy and particles that are constantly popping in and out of existence.

In 1948, the effect was predicted by Dutch physicist Hendrick Casimir, who realized that close metal plates would block out light waves that are too big to fit between them. If the gap was only a few nanometres (billionths of a metre) wide, the energy density outside the plates would be higher than between the plates, creating a pressure to push them together. In a nautical analogy, two large ships that are side by side in windless conditions drift together. The ships cancel waves between them, while waves outside buffet the ships together.

The Casimir effect can also be a repulsive force, depending on the experimental arrangement. It could one day be useful in nanoscale machinery (see page 122), creating repulsion between components that lets them move without friction.

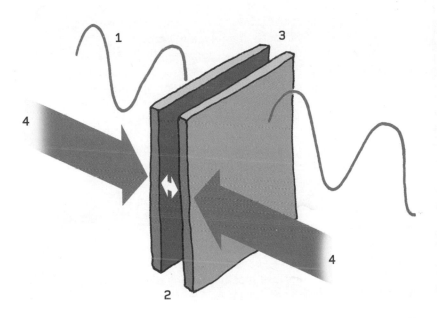

1 Long wavelengths of light in space around plates

2 Plates are separated by a fraction of the wavelength of light

3 Light cannot pass between closely separated plates

4 External pressure from energy density difference pushes plates together

Superfluids

Superfluids are fluids that have no viscosity, or stickiness, and so move without experiencing friction. In 1962, experiments produced the first superfluid using helium-4, chilled to just 2.17°C (3.91°F) above absolute zero. Helium-3 also forms a superfluid, but at an even lower temperature.

Superfluids are famous for their weird behaviour. Placed in a beaker, superfluid helium creeps up the sides and over the top. Another strange property is that its spin is quantized – it only rotates at certain allowed speeds. When the container of a superfluid rotates below the liquid's sound speed, the fluid doesn't move. Once the container's speed reaches the sound speed, the superfluid suddenly spins at this speed.

Perfect superfluids also have infinite thermal conductivity. A hot spot in superfluid helium will ripple through it like a sound wave with a speed of around 20 m/s (45 mph). The name superfluid was coined to echo the term superconductor (see page 82), which describes materials that conduct electric currents without any resistance.

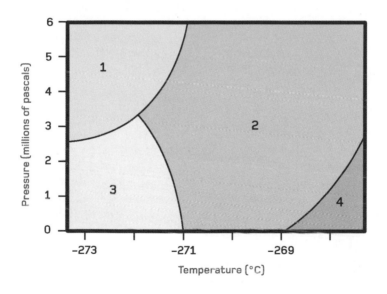

Helium–4 phases at low temperature

1 Solid 3 Superfluid
2 Normal liquid 4 Gas

Bose–Einstein condensates

A Bose–Einstein condensate is a peculiar state of matter that some particles form when they all crash down into their lowest possible energy states at temperatures close to absolute zero. The condensates are an interesting window on the physics of quantum mechanics, because they illustrate quantum effects on a macroscopic scale.

In the mid-1920s, Indian physicist Satyendra Nath Bose and Albert Einstein predicted the existence of these condensates. They form from particles called bosons, which have integer (whole-number) values of a quantum property called spin.

In 1995, scientists in Colorado produced the first Bose–Einstein condensate by chilling rubidium atoms to nearly absolute zero. The atoms overlap to form a blob that behaves like a single 'superatom'. Bose–Einstein condensates might one day have a practical use – lasers have flourished in technologies because they create identical light photons that are easy to control; likewise, Bose–Einstein condensates could fuel technologies that need precise control of identical atoms.

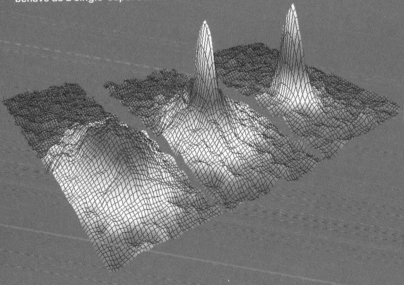

A computer model shows the emergence (from left to right) of a Bose–Einstein condensate as individual atoms begin to take on identical quantum properties and behave as a single 'superatom'.

Superconductivity

A superconductor is a material that can conduct electricity without any resistance. Once set in motion, a current will flow forever in a closed loop of superconducting material.

Superconductivity was first discovered in the element mercury. At a temperature of just 4°C (7.2°F) above absolute zero, its electrical resistance switches off. Theory suggests low-temperature superconductivity arises because electrons passing through a crystal lattice deform the lattice, creating 'troughs' of positive charge that help propel subsequent electrons through the same region.

Known superconductors include metals, polymers and even ceramics. Superconducting coils cooled to very low temperatures are used as superconducting magnets, which can produce extremely strong magnetic fields. They are used in medical scanners and levitating 'maglev' trains, which have achieved speeds of more than 580 km/h (360 mph). The holy grail is to find materials that are superconducting at easily achievable temperatures above 0°C (32°F).

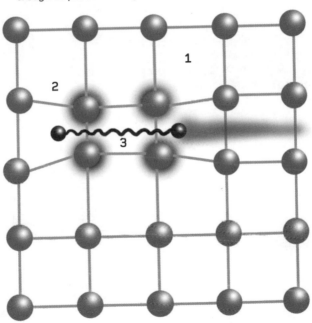

1 Crystalline lattice of conducting material

2 Deformed lattice forms 'trough' of positive charge

3 Electrons lock together in 'Cooper pairs' that slip through the lattice easily

Standard model particles

The standard model describes the most fundamental particles in nature. The tiniest constituents of matter form two families. The first are the quarks, divided into six 'flavours' called up, down, charm, strange, top and bottom. These feel the strong force (see page 86) and carry an electric charge of $+\frac{2}{3}$ or $-\frac{1}{3}$ times that of the electron. Quarks combine in pairs or threes to form other particles, such as protons and neutrons.

The second group of matter particles are leptons, which don't feel the strong force. The most familiar lepton is the electron, which has two heavier siblings called the muon and tau. All three have the same electric charge, -1. The final three leptons are 'neutrinos', electrically neutral particles with tiny masses. These stream out from nuclear reactions in the Sun and easily fly straight through the Earth.

The standard model also includes several force-transmitting particles called gauge bosons, including the photon (see page 54). Scientists suspect there is a 'Higgs boson' that endows fundamental particles with mass, but so far it remains elusive.

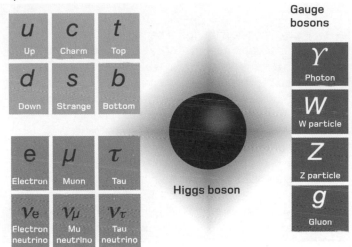

Quarks

u Up	*c* Charm	*t* Top
d Down	*s* Strange	*b* Bottom

e Electron	*μ* Muon	*τ* Tau
ν_e Electron neutrino	ν_μ Mu neutrino	ν_τ Tau neutrino

Leptons

Gauge bosons

Y Photon
W W particle
Z Z particle
g Gluon

Higgs boson

Strong and weak forces

In particle physics, the strong force (also called the strong interaction or strong nuclear force) is a fundamental force of nature. It binds quarks together to form protons and neutrons (see page 84), and also binds protons and neutrons together inside atomic nuclei. The strong force has a range similar to the size of a typical atomic nucleus.

The weak force, or weak interaction, is another fundamental force of nature. Its range is tiny, only about a thousandth of the size of a proton. Its most familiar effect is beta decay, in which it allows a nucleus to emit an electron or positron and change its overall electric charge. The weak force also initiates hydrogen fusion in stars, and allows one quark to change into another 'flavour' of quark.

Scientists hope to eventually craft a single 'theory of everything' that elegantly describes the behaviour of all four forces – strong, weak, electromagnetic and gravitational – within the same mathematical framework (see page 90).

The strong force binds protons (1) and neutrons (2) in nuclei, such as those shown here, as well as the 'up' (3) and 'down' (4) quarks within individual protons and neutrons.

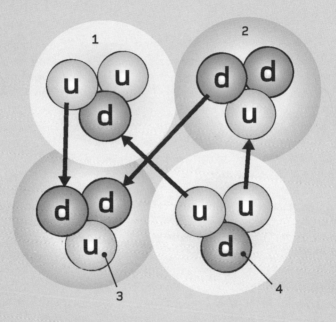

Antimatter

In a nutshell, antimatter is matter's nemesis. Every matter particle from the standard model (see page 84) has an antimatter counterpart with equal mass but opposite charge, and when the two meet, they destroy one another on contact.

The British physicist Paul Dirac predicted in the 1920s that there must be a particle in nature that is identical to the electron but with opposite charge. This anti-electron or 'positron' was discovered experimentally in 1932. When an electron meets a positron, they annihilate one another, disappearing in a puff of gamma rays. Futuristic scenarios for interstellar travel propose using antimatter as fuel, because its reaction with matter releases energy so efficiently.

One puzzle about antimatter remains. Theory suggests that the newborn universe had equal amounts of matter and antimatter, so why is matter so dominant today? Possibly, some slight asymmetry allowed matter to win out. A more exotic possibility is that distant antimatter realms in the universe survive to this day, teeming with galaxies of antimatter stars.

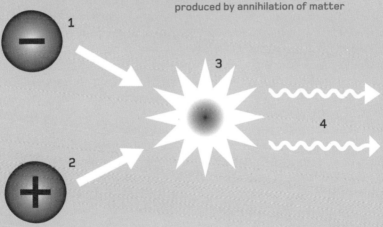

1 Electron
2 Anti-electron or 'positron'
3 When electron and positron meet, an annihilation event occurs
4 High-energy gamma rays produced by annihilation of matter

Grand unified theories

Grand unified theories try to mathematically describe the forces of nature under one umbrella. Theory has unified the electromagnetic force and the weak force (see page 86), showing that they acted like a single force in the hot, early universe, when particles were highly energetic. But to date, no satisfactory theory unites them with the strong force as well.

A satisfactory grand unified theory would explain various aspects of the standard model particles (see page 84) and forces that so far remain a mystery. For instance, why are there six quarks and six leptons? Why do they have their particular masses? But the grand unified theories developed so far are unpleasantly complicated and invoke exotic, untested physics, some requiring space to have hidden extra dimensions.

The ultimate goal is to also unify gravity with the other forces in a 'theory of everything'. One candidate is string theory, which assumes that particles are like tiny vibrating strings. But string theory has yet to make testable predictions to prove that it accurately describes nature's design.

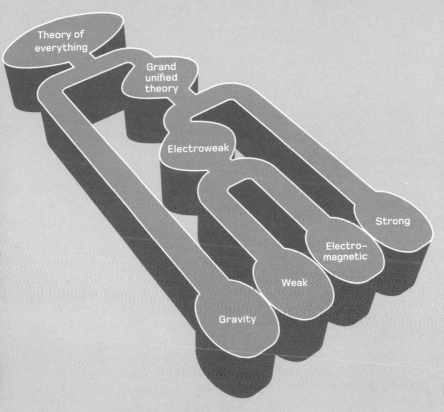

Atomic structure

Atoms consist of a tiny dense nucleus containing positively charged protons and uncharged neutrons, surrounded by clouds of electrons. Because protons and neutrons are much heavier than electrons, most of an atom's mass resides in the central nucleus.

Each chemical element (see page 100) has a unique number of protons in its nucleus, its 'atomic number'. For instance, the element carbon with six protons has the atomic number six. However, single elements can have different numbers of neutrons in the nucleus. For example, carbon has three naturally occurring 'isotopes' (see page 102) with six, seven or eight neutrons. The sum of protons and neutrons in an atom's nucleus is called the atomic mass number.

Normally, the net electric charge of an atom is zero because the number of electrons is the same as the number of protons, and their equal and opposite electric charges cancel out. However, it's possible to knock electrons out of atoms or add extra ones to create positively or negatively charged 'ions'.

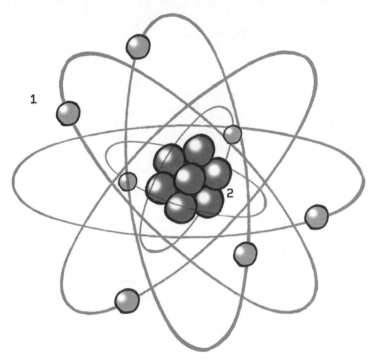

1 Negatively charged electrons in orbit around nucleus
2 Nucleus containing positively charged protons and neutrons

Atomic nucleus

The atomic nucleus is the dense cluster of protons and neutrons that sits at the centre of an atom, surrounded by clouds of electrons. Nuclei are tiny compared to atoms themselves. If an atom was scaled up to the size of a football stadium, its nucleus would typically be about the size of a pea.

The atom's structure was unclear until 1909, when a famous experiment by New Zealand physicist Ernest Rutherford showed that positive charge is densely concentrated in the middle. His team fired positively charged alpha particles at a thin sheet of gold foil and found that most flew through the foil in a straight line. However, a tiny number bounced off at large angles. Rutherford realized these had happened to hit a tiny, positively charged nucleus at an atom's centre.

Atomic nuclei are now known to be made up of protons and neutrons. The protons feel repulsion from each other due to their positive electric charge, but the attractive strong force between the protons and neutrons overcomes this repulsion to hold nuclei together.

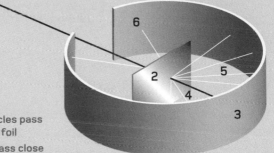

The Rutherford Experiment

1 Alpha-particle source
2 Thin gold foil
3 Radiation detector

4 Most alpha particles pass straight through foil

5 A small number pass close to an atomic nucleus and are deflected

6 Occasionally, an alpha particle collides directly with an atomic nucleus and rebounds

Radioactivity

Radioactivity involves the spontaneous decomposition of an unstable atomic nucleus into a more stable type. There are three main types of decay, which were named simply 'alpha', 'beta' and 'gamma' in the days when they were poorly understood, and these names have stuck.

Alpha decay occurs when a heavy nucleus emits a particle containing two protons and two neutrons. Uranium-238 transforms into thorium-234, for instance, which has two fewer protons and two fewer neutrons. In beta decay, a neutron can convert into a proton with the emission of an electron, increasing the atomic number by one. Alternatively, an excited nucleus can emit a gamma ray.

Lead is the heaviest stable element, all heavier ones decaying over time. Radioactivity is a random, unpredictable process but the decay rates of many identical atoms can be reliably characterized by a 'half life', the time it takes for half the nuclei to decay. Half lives vary from a fraction of a second to many billions of years – longer even than the age of the universe.

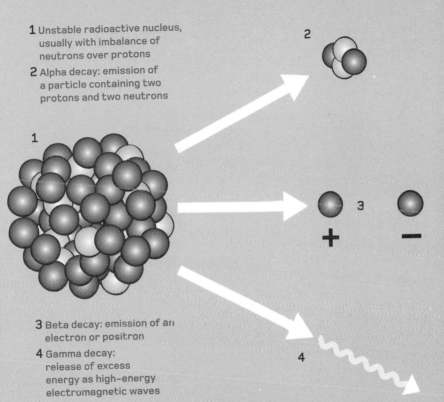

1 Unstable radioactive nucleus, usually with imbalance of neutrons over protons

2 Alpha decay: emission of a particle containing two protons and two neutrons

3 Beta decay: emission of an electron or positron

4 Gamma decay: release of excess energy as high-energy electromagnetic waves

Nuclear fission and fusion

Nuclear fission occurs when a heavy atomic nucleus splits into two, with the release of energy. Nuclei are made up of protons and neutrons, but the mass of a nucleus is always less than the sum of the individual masses of the protons and neutrons inside. The difference is a measure of the 'nuclear binding energy' that holds the nucleus together, and this energy is released when the nucleus splits. For example, uranium-235 can split to form two lighter elements such as rubidium and caesium.

Nuclear fusion is the opposite process, in which two light nuclei merge to form a heavier one, yielding energy because the combination is lighter than the sum of its parts. Atoms heavier than iron can undergo fission, while lighter ones can fuse.

Nuclear power stations generate energy from fission reactions. About 2 billion years ago, natural fission took place at Oklo in Gabon, Africa, when groundwater concentrated uranium deposits. Fusion occurs in the Sun's core, where hydrogen nuclei fuse into helium, generating the Sun's energy.

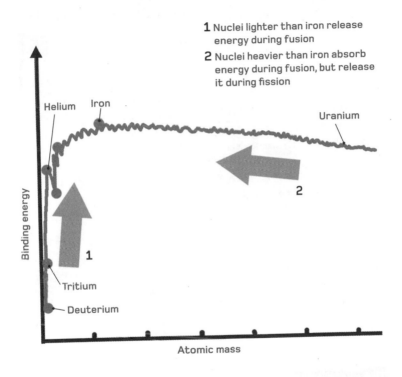

1 Nuclei lighter than iron release energy during fusion

2 Nuclei heavier than iron absorb energy during fusion, but release it during fission

Chemical elements

Chemical elements are the simplest substances found in nature, consisting of individual atoms that all have the same number of protons (the 'atomic number') in the atomic nucleus. Each nucleus is surrounded by shells of negatively charged electrons that usually cancel out the positive nuclear charge, making the atom neutral overall.

Each element has a standard chemical symbol, such as H for hydrogen and Fe for iron. Hydrogen is the lightest element, consisting of just one proton and one electron, while elements heavier than uranium, which has atomic number 92, are all unstable and rapidly undergo radioactive decay (see page 96).

The periodic table displays chemical elements in rows that highlight repeating trends. Atomic number increases from left to right, while the chemical properties of elements in each column are similar. For instance, the far right column contains neon and argon, inert gases that don't easily form compounds. That's because they have the same configurations of outer electrons, the key factor in chemical properties.

The Periodic Table

1																	18
1 H	2											13	14	15	16	17	2 He
3 Li	4 Be											5 B	6 C	7 N	8 O	9 F	10 Ne
11 Na	12 Mg	3	4	5	6	7	8	9	10	11	12	13 Al	14 Si	15 P	16 S	17 Cl	18 Ar
19 K	20 Ca	21 Sc	22 Ti	23 V	24 Cr	25 Mn	26 Fe	27 Co	28 Ni	29 Cu	30 Zn	31 Ga	32 Ge	33 As	34 Se	35 Br	36 Kr
37 Rb	38 Sr	39 Y	40 Zr	41 Nb	42 Mo	43 Tc	44 Ru	45 Rh	46 Pd	47 Ag	48 Cd	49 In	50 Sn	51 Sb	52 Te	53 I	54 Xe
55 Cs	56 Ba		72 Hf	73 Ta	74 W	75 Re	76 Os	77 Ir	78 Pt	79 Au	80 Hg	81 Tl	82 Pb	83 Bi	84 Po	85 At	86 Rn
87 Fr	88 Ra		104 Rf	105 Db	106 Sg	107 Bh	108 Hs	109 Mt	110 Ds	111 Rg	112 Uub	113 Uut	114 Uuq	115 Uup	116 Uuh	117 Uus	118 Uuo

57 La	58 Ce	59 Pr	60 Nd	61 Pm	62 Sm	63 Eu	64 Gd	65 Tb	66 Dy	67 Ho	68 Er	69 Tm	70 Yb	71 Lu
89 Ac	90 Th	91 Pa	92 U	93 Np	94 Pu	95 Am	96 Cm	97 Bk	98 Cf	99 Es	100 Fm	101 Md	102 No	103 Lr

Isotopes

Chemical elements can exist as two or more isotopes that have different numbers of neutrons in the nucleus. For example, while carbon always has six nuclear protons, it exists as three different naturally occurring isotopes with six, seven or eight neutrons. These isotopes are often written carbon-12, carbon-13 and carbon-14.

Chemically, different isotopes of an element are usually identical because their chemical properties are determined by their outer electrons. But different isotopes undergo nuclear decay at different rates. For instance, while most carbon on Earth is the stable isotope carbon-12, the isotope carbon-14 is radioactive and decays with a half-life of 5,700 years.

This underpins the technique of carbon dating. Constant interchange with the environment makes the ratio of carbon-14 to carbon-12 constant in a living tree, for instance, but the ratio drops with time in a predictable way after the tree dies. If ancient wood has just half the expected 'living' value of carbon-14, it must be about 5,700 years old.

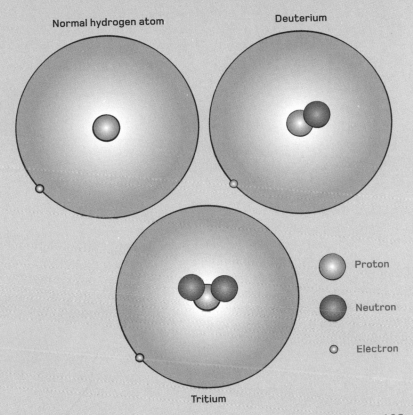

Normal hydrogen atom

Deuterium

Tritium

Proton

Neutron

Electron

CHEMISTRY 103

Allotropes

The atoms of some elements sometimes bind together into different structures called allotropes. For example, oxygen in the Earth's atmosphere exists as two allotropes – stable diatomic oxygen (O_2) and ozone (O_3). Ozone is an unstable molecule that forms when diatomic oxygen absorbs ultraviolet light from the Sun.

Solid carbon has three common allotropes. Diamond consists of carbon atoms bonded in a tetrahedral lattice, while graphite consists of flat sheets of carbon atoms bound in hexagons. Fullerenes are spheres ('buckyballs') or tubes of carbon atoms, including the soccer-ball-shaped molecule C_{60}.

Allotropes of an element can have very different physical and chemical properties. Diamond is the hardest known mineral in nature because each carbon atom is bonded rigidly to four other carbons in a tetrahedron; graphite is relatively soft because the flat sheets are weakly bonded and can slide over each other. While normal diatomic oxygen forms a colourless, odourless gas, ozone is a pale blue gas with a pungent smell.

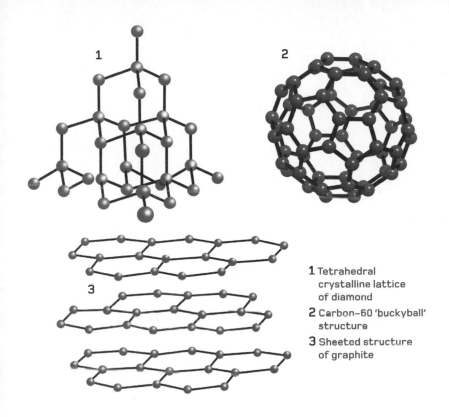

1 Tetrahedral crystalline lattice of diamond

2 Carbon-60 'buckyball' structure

3 Sheeted structure of graphite

Solutions and compounds

Atoms of different elements can join together in chemical reactions to form compounds. For example, the elements hydrogen and oxygen react to form water (H_2O). The properties of compounds are usually very different from the properties of the elements they contain; hydrogen and oxygen are gases at room temperature, for instance, but water is liquid.

Compounds always have a fixed ratio of atoms that are held together in set arrangements by chemical bonds, and they can only be separated into elements by chemical reactions. Unlike compounds, mixtures consist of two or more substances that do not combine chemically and can usually be separated by simple mechanical means such as filtering or evaporation.

Mixtures include alloys, such as steel (iron with carbon), as well as solutions, such as salt dissolved in water. 'Colloids' are substances with particles evenly dispersed throughout them, such as emulsion paint, while suspensions are fluids containing solid particles that are large enough to gradually settle out of the fluid.

1 A dilute copper sulphate solution contains relatively few dissolved copper sulphate molecules (the solute) compared to the amount of water (the solvent)

2 When more copper sulphate is added, the solution becomes increasingly concentrated

3 Eventually the solvent can hold no more solute, and is said to be saturated

Chemical bonds

Chemical bonds bind elements together to form compounds. Chemical bonding occurs because atoms are most stable when their outer electron shell – also called the valence shell – is either completely full or empty.

Covalent bonds form when atoms team up to fill their valence shells by sharing their outer electrons. For instance, atoms of hydrogen have only one valence electron in a shell that can hold a maximum of two electrons. Hydrogen molecules form because two hydrogen atoms join up to share their outer electrons and attain full valence shells. Oxygen has two electron vacancies in its valence shell, so it covalently bonds to two hydrogen atoms to form water.

Ionic bonding occurs when a substance, usually a metal, donates an electron to another atom. For instance, sodium chloride (common salt) forms when sodium donates an electron to chlorine. The sodium and chloride ions then have opposite electrical charge and the electrostatic force between them holds the molecule together.

Molecular hydrogen (H₂)

Atoms each share an electron in their outer shell, filling their valence shell requirement of two electrons and forming a covalent bond

Oxygen shares electrons with two hydrogen atoms, achieving a full shell of eight electrons to form a covalent bond

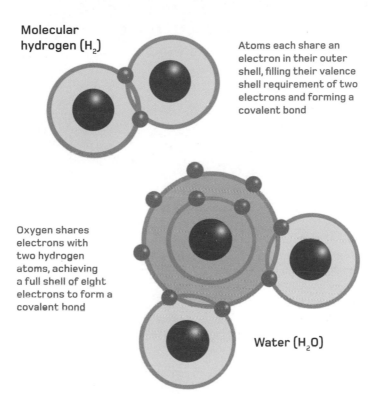

Water (H₂O)

Chemical reactions

In chemistry, a reaction happens when two or more atoms or molecules interact and transform into a different compound. For example, rusting occurs when an 'oxidation' reaction makes iron combine with oxygen to form rust-coloured iron oxide.

The opposite process, such as reactions that remove oxygen from iron ores like haematite (Fe_2O_3), is called reduction. More generally, oxidation means an atom loses electrons as it bonds, while it gains electrons during reduction. Combustion or burning involves reactions between a fuel and an oxidant, with the release of heat. For example, methane or natural gas burns in oxygen to form water vapour and carbon dioxide.

'Catalysts' are substances that can increase the rate of a chemical reaction, without chemically changing themselves. Some reactions are reversible, such as the 'Haber process' in which nitrogen and hydrogen combine to form ammonia (NH_3). The reaction is said to be in equilibrium when the forward reaction occurs at the same rate as the reverse reaction, which breaks ammonia back down into nitrogen and hydrogen.

Combustion of methane

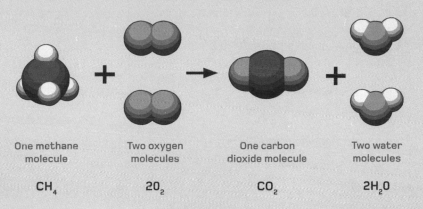

| One methane molecule | Two oxygen molecules | One carbon dioxide molecule | Two water molecules |

CH_4 $2O_2$ CO_2 $2H_2O$

Acids and bases

Generally speaking, acids are solutions containing an excess of positive hydrogen ions, while bases or alkaline solutions contain an excess of negatively charged hydroxide (OH^-) ions. Acids and bases also have more general definitions as electron acceptors and electron donors.

An example of an acid is hydrochloric acid, which forms when hydrogen chloride (H^+Cl^-) dissolves in water and the

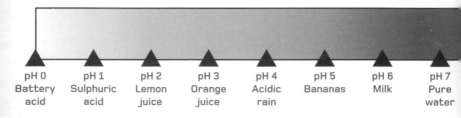

| pH 0 Battery acid | pH 1 Sulphuric acid | pH 2 Lemon juice | pH 3 Orange juice | pH 4 Acidic rain | pH 5 Bananas | pH 6 Milk | pH 7 Pure water |

bonds between the hydrogen and chloride ions break, liberating free positive hydrogen ions. Likewise, dissolved sodium hydroxide (Na^+OH^-) creates an alkaline solution. The pH scale measures acidity and ranges from 0 (highly acidic) to 14 (highly alkaline). Car battery acid has a pH of around 0 to 1, while milk of magnesia has a pH around 10. Perfectly pure water has a neutral pH of 7.

Acids and bases neutralize each other because excess hydrogen ions combine with excess hydroxide ions to form water. These neutralization reactions also form various salts. For example, hydrochloric acid reacts with sodium hydroxide to produce water and sodium chloride, common table salt.

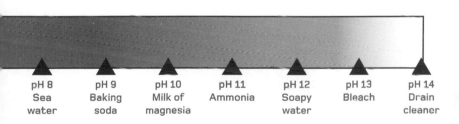

| pH 8 | pH 9 | pH 10 | pH 11 | pH 12 | pH 13 | pH 14 |
| Sea water | Baking soda | Milk of magnesia | Ammonia | Soapy water | Bleach | Drain cleaner |

Electrolysis

Electrolysis is a process that uses electricity to drive a chemical reaction. When positive and negative electrodes are placed in a fluid, positively charged ions in the fluid drift towards the negative electrode, where they accept electrons, while negatively charged ions move to the positive electrode, where they are oxidized. For example, molten aluminium oxide can be electrolysed to produce pure aluminium at the negative electrode, while oxygen bubbles off at the positive electrode.

Batteries effectively reverse this process, with chemical reactions generating electrical energy. When plates of copper and zinc are placed in a sulphuric acid solution, current flows between them. The zinc electrode gives up electrons that flow along a wire to a copper plate. Once there, they combine with hydrogen ions to liberate hydrogen gas. Many modern batteries use a paste of potassium hydroxide as their electrolyte.

Fuel cells are like batteries, but consume fuel from an outside source. For example, they can generate electricity by oxidizing a constant feed of hydrogen gas to form water.

A simple battery

1 Zinc anode gives
   up electrons
2 Sulphuric acid
   electrolyte
3 Copper cathode
   receives electrons
4 Electric current
   flows through wire

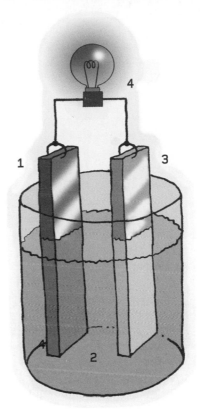

Molecular geometry

Molecular geometry describes the overall shape of a molecule in terms of how the atoms inside it are arranged. Examples of simple structures are linear molecules like carbon dioxide ($O=C=O$) and tetrahedral molecules like methane, which consists of a carbon atom with four hydrogen atoms surrounding it at the corners of a tetrahedron.

Trigonal-bipyramidal molecules are shaped like two pyramids back to back, while octahedral molecules have a shape like an eight-sided solid. Octahedral molecules include the compound sulphur hexafluoride (SF_6).

'Isomers' are compounds that have the same chemical formula but different molecular structures. For instance, the sugar fructose is an isomer of glucose – they have the same formula $C_6H_{12}O_6$, but their atoms are arranged in different ways. Sometimes, two isomers are mirror images of each other, in which case the molecule is said to be 'chiral' and the two mirror-image forms are called enantiomers. Chiral molecules include most amino acids, the building blocks of proteins.

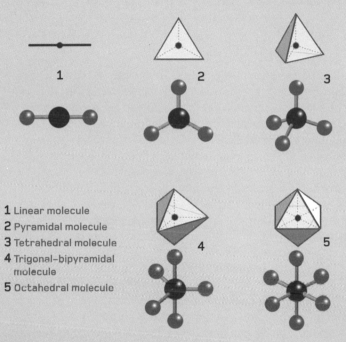

1 Linear molecule
2 Pyramidal molecule
3 Tetrahedral molecule
4 Trigonal-bipyramidal
  molecule
5 Octahedral molecule

Structural formulas

The structural formula of a molecule indicates how the atoms inside it are bound together. For example, ethanol has the chemical formula C_2H_6O, but its structural formula is CH_3-CH_2-OH, indicating that a methyl group (CH_3) is attached to the carbon of a methylene group (CH_2), which in turn is attached to the oxygen of a hydroxyl group (OH).

There are various graphical ways of representing a structural formula, including the simple flat 'Lewis structure' that shows how atoms bond together. A representation called the Natta projection shows molecules in three dimensions, with solid and dotted triangular bonds indicating bonding directions towards and away from the viewer respectively.

Often, 'skeletal formulas' are used to describe complex organic molecules, with a hexagon representing the benzene ring C_6H_6, for instance. To keep things simple, skeletal formulas don't label carbon and hydrogen atoms specifically – carbon is assumed to sit at the vertices with as much hydrogen as it needs to use up four bonds.

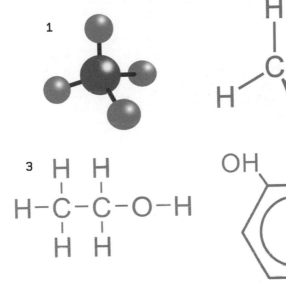

1 Three-dimensional model of methane
2 Natta projection of methane
3 Lewis structure of ethanol
4 Skeletal formula of hydroquinone

Chemical polarity

Polar molecules are ones in which the electric charge is unevenly distributed, so that one side of the molecule is positively charged while another region is negatively charged.

Water is an example of a polar molecule. There is an excess of positive charge on the side of the molecule where two hydrogen atoms sit, covalently bonded to oxygen via a shared pair of electrons. The oxygen also has two unshared electron pairs on the opposite side of the hydrogen atoms, making that side negatively charged.

Water molecules tend to align themselves so that the negatively charged side of one molecule sits next to the positively charged end of an adjacent one. This creates a weak type of secondary bonding called hydrogen bonding. These bonds give water a crystal structure when it freezes, and this explains why water ice is less dense than liquid water. As a result, when ice forms on a lake during a cold winter, it floats to the surface, forming an insulating blanket that protects the entire lake from freezing.

Polarity of a water molecule

1 Zone of negative polarity around unbonded side of oxygen atom

2 Zones of positive polarity around hydrogen atoms

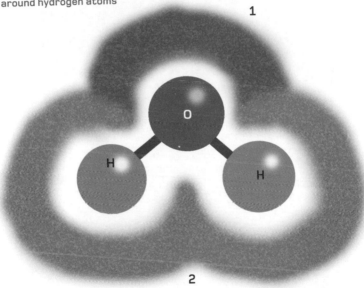

Molecular engineering

Molecular engineering, or nanotechnology, is the manipulation of matter on tiny scales down to a billionth of a metre (about one hundred-thousandth of the width of a human hair). It creates materials with useful nanoscale properties, including invisible coatings just 3 micrometres (millionths of a metre) thick that protect shiny stainless steel exhaust pipes on cars from corrosion.

Opticians apply nanocoatings to glasses to make them more scratch resistant and easier to keep clean, while other nanomaterials help strengthen composite materials used in lightweight tennis rackets and bicycles, for instance. However, some scientists worry that nanoparticles in commercial products could cause diseases such as lung cancer if inhaled.

Nanomachines or 'nanobots' are in an early research and development phase. In future, tiny nanosensors in packaging could detect the pathogens that cause food poisoning, while nanobots could swim through your bloodstream to repair DNA damage in cells, or identify and kill tumours.

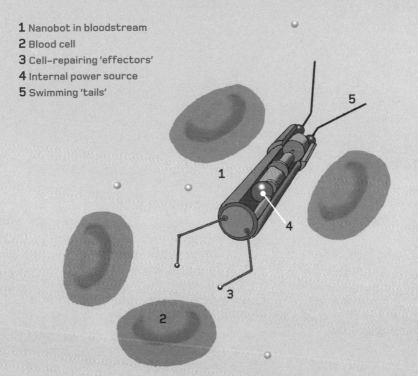

1 Nanobot in bloodstream
2 Blood cell
3 Cell-repairing 'effectors'
4 Internal power source
5 Swimming 'tails'

Crystal structures

A crystal or crystalline solid is a material in which the atoms or molecules are arranged in a rigid and orderly repeating pattern. Table salt, snowflakes and diamonds are common examples of crystals. Crystalline rocks can form in solutions or when molten magma cools. For instance, completely crystallized granite forms when magma cools and solidifies very slowly under high pressure.

Crystals can have a simple cubic lattice, with one lattice point on each

1

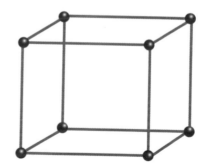

1 Simple cubic lattice
2 Body-centred cubic lattice
3 Face-centred cubic lattice

corner of a cube, while the body-centred cubic system also has a lattice point in the centre of the cube. The face-centred cubic system has lattice points in the middle of the cube faces. Common salt forms a face-centred cubic lattice with alternating atoms of sodium and chlorine.

Some crystals can also form more complicated shapes, including double pyramids and eight-faced octahedra. Scientists often study the structures of crystals by passing X-rays through them and examining the resulting diffraction patterns (see page 60).

2

3

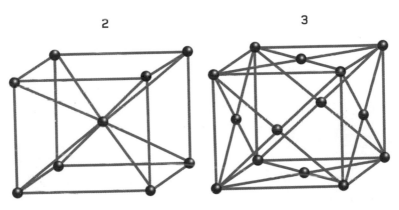

Metals

In chemistry, a metal is an element or alloy that has high electrical and thermal conductivity. A metal's ability to conduct electricity and heat stems from the fact that its outer electrons are extremely loosely bound to the atoms and readily flow through a metal wire. Iron and aluminium are the two most common metals on Earth.

Metals are typically denser than non-metallic elements and they readily form positively charged ions by losing electrons, although their levels of reactivity vary as shown by the diagram opposite – iron rusts over years as it converts to iron oxide in the atmosphere, while pure potassium burns up in seconds as it oxidizes. Some metals, such as the precious metals platinum and gold, do not react with the atmosphere at all. Others, including aluminium and titanium, form a thin oxide layer on their surfaces that protects them from further oxidation.

Confusingly, however, astronomers often use the term 'metal' to refer to *any* element in the universe that is heavier than hydrogen or helium.

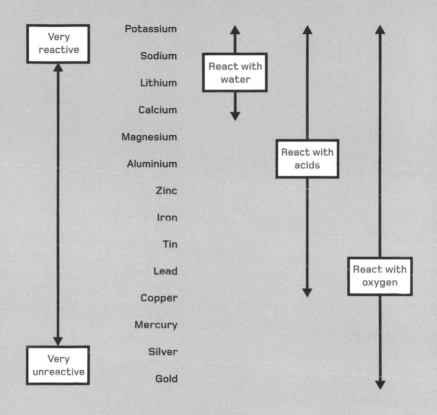

Semiconductors

A semiconductor is a material that conducts electricity better than an insulator (including most ceramics) but less well than an electrical conductor like copper. Semiconductors can be pure elements, such as silicon or germanium, or compounds, including gallium arsenide or cadmium selenide.

When electrons move in a semiconductor, they leave behind 'holes' with relative positive charge, and this makes semiconductors useful for electronic devices such as transistors, often used as switches. An example is the NPN transistor, which sandwiches a P-type semiconductor (with an excess of positive holes) between two N-type semiconductors (with an excess of negatively charged electrons).

When an electric current is applied to the 'base' input of the transistor, it increases the conductivity of the P-type region, which in turn increases current flow across the transistor from the 'collector' to the 'emitter'. Today, transistors are miniaturized on microchips (see page 374). Semiconductors play vital roles in just about all modern electronic devices.

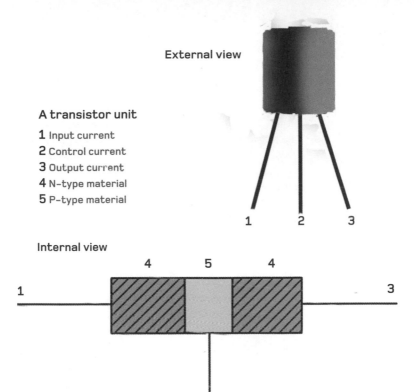

External view

A transistor unit

1 Input current
2 Control current
3 Output current
4 N-type material
5 P-type material

Internal view

Polymers

Polymers are materials made from large molecules composed of many repeating units. Naturally occurring polymers include starch, which is made up of many repeating units of the sugar glucose, and proteins, which are long chains of amino acids. Most polymers are organic, using carbon bonds as their backbone.

Plastics are synthetic polymers. Polyethylene, commonly called polythene, is one of the simplest, composed of chains of repeating CH_2 (ethylene) units. Depending on the temperatures and pressures that are applied during production, the ethylene molecules can link up to create high-density polyethylene, used in containers like milk bottles, or low-density polyethylene, used to produce plastic films and sandwich bags, for instance.

Polyvinyl chloride is a polymer similar to polyethylene, but also includes chlorine atoms. This rigid polymer is used for pipes, window frames and vinyl siding for houses. It can be blended with other compounds to make it soft for products like raincoats and shower curtains.

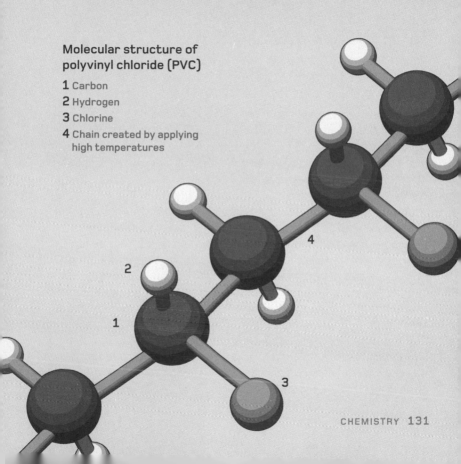

Molecular structure of
polyvinyl chloride (PVC)

1 Carbon
2 Hydrogen
3 Chlorine
4 Chain created by applying
high temperatures

Composites

Composite materials are ones that combine two or more materials without completely blending them. An example is reinforced concrete, which uses cement to bind rough gravel and sand, often with steel bars running through it for extra strength. Wood is a natural composite, composed of cellulose fibres in a matrix of the complex polymer lignin.

Usually, composite materials are designed to be lightweight yet strong and stiff. Typically, one material (the matrix or binder) surrounds and binds together clusters of fibres of a much stronger material (the reinforcement). For instance, fibreglass is a plastic matrix reinforced by threads of glass.

When building an aircraft, engineers need lightweight, strong materials that can withstand the stresses from turbulent air. One material they use is carbon-fibre-reinforced plastic, which is similar to fibreglass but even stronger. Aerospace engineers often build spacecraft with more exotic composite materials designed to withstand extremely low temperatures in Earth orbit or interplanetary space.

Materials used in a typical jet airliner

- ■ Aluminium/steel composite
- □ Aluminium
- ■ Carbon laminate composite
- ■ Carbon sandwich composite

Nanomaterials

Nanomaterials are substances that have at least one dimension smaller than about 100 nanometres (billionths of a metre), which is about one-thousandth of the width of a human hair. Nanomaterials can be nanoscale in just one dimension (such as surface films), two dimensions (fibres or strands) or three dimensions (tiny particles).

Many nanomaterials have unusual properties because they reach the ultra-small realms where the quantum behaviours of atoms start to show their hand. Products containing nanomaterials are already in commercial use, including sunscreens with nanoparticles that absorb ultraviolet sunlight without generating skin-damaging free radicals in the process. Other products include stain-resistant textiles.

A nanoscale coating of titanium dioxide makes windows 'self-cleaning'. When the coating absorbs UV sunlight, it breaks down organic dirt. The coating is also 'hydrophilic', or water loving, so that rain forms a sheet on the glass rather than individual droplets, keeping the glass evenly clean.

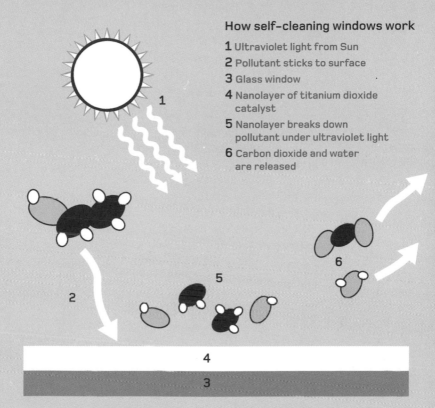

How self-cleaning windows work

1 Ultraviolet light from Sun
2 Pollutant sticks to surface
3 Glass window
4 Nanolayer of titanium dioxide catalyst
5 Nanolayer breaks down pollutant under ultraviolet light
6 Carbon dioxide and water are released

Metamaterials

Metamaterials are materials that have been artificially engineered to have exotic properties unlike anything we've seen in nature. One example is material with the potential to act like the 'cloaking' devices that make spaceships or people invisible in sci-fi films.

In this case, materials are carefully engineered to manipulate light. Sometimes they alternate tiny layers to produce a substance with a negative refractive index (see page 58) that bends light in unexpected directions. In theory, such a material could cloak an object by making light waves pass around it and then continue along their original straight paths, so the object would be invisible to someone standing 'downstream'.

Research is at a very early stage and is unlikely to lead to viable invisibility cloaks. However, similar materials could be extremely useful in future microscopes for imaging tiny viruses and molecules, because they don't suffer from the diffraction limit that prevents normal materials focusing light to a tiny super-sharp spot (see page 60).

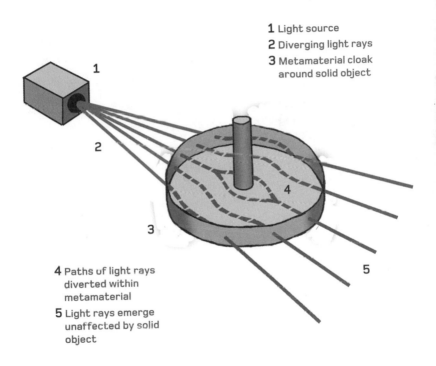

1 Light source

2 Diverging light rays

3 Metamaterial cloak around solid object

4 Paths of light rays diverted within metamaterial

5 Light rays emerge unaffected by solid object

Proteins

Proteins are large, complex molecules that play many critical roles in cells. They consist of long chains of hundreds or thousands of smaller simple molecules called amino acids. Some of these are 'essential amino acids', such as phenylalanine, which our bodies don't synthesize, so it's essential that we eat plenty of these in our diet.

Some proteins act as antibodies, which can prevent disease by targeting foreign particles such as viruses and blocking the sites they use to invade cells. Others include receptors (see page 148) and enzymes, which enable thousands of chemical reactions to take place in cells and assist the construction of new proteins by reading the genetic information in DNA.

In plants and animals, all proteins are made from different sequences of 20 main amino acids. The exact sequence for a given protein is called its primary structure. When a cell makes a new protein, it forms a linear chain of amino acids that coils into a secondary structure before morphing into a final three-dimensional shape through a process called protein folding.

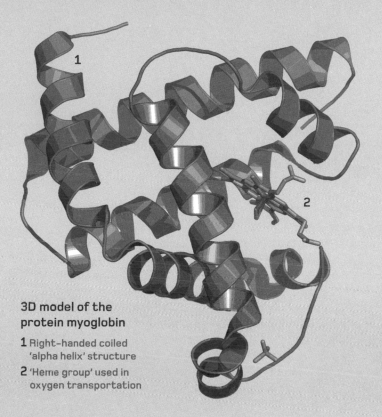

3D model of the protein myoglobin

1 Right-handed coiled 'alpha helix' structure

2 'Heme group' used in oxygen transportation

Carbohydrates

Carbohydrates are organic compounds made of carbon, hydrogen and oxygen atoms. In food science and informal contexts, the term 'carbohydrate' often just means any sugary food like chocolate or a starchy food such as bread or pasta.

The most basic carbohydrates are simple sugars or 'monosaccharides' such as fructose ($C_6H_{12}O_6$), which makes fruit sweet, or ribose ($C_5H_{10}O_5$), which forms the backbone of the genetic molecule RNA. Monosaccharides, especially glucose, are also the major source of fuel for metabolism (see page 144). Glucose has the same chemical formula as fructose but a different, typically ring-shaped, molecular structure.

The larger 'disaccharide' sucrose ($C_{12}H_{22}O_{11}$) – common table sugar – is formed from fructose and glucose. The most complicated carbohydrates are 'polysaccharides', including starch, which is made up of thousands of glucose units. Plants store their glucose fuel as starch. Many animals, including humans, store glucose as glycogen, a molecule containing a core protein surrounded by many branching glucose units.

Structure of the carbohydrate
D-Glucose (α-ring form)

1 Carbon
2 Hydrogen
3 Oxygen

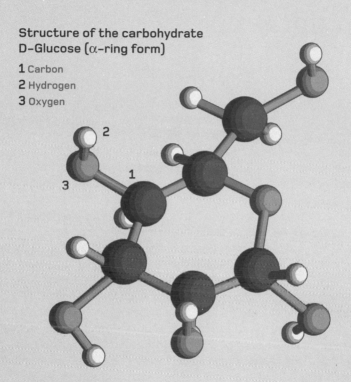

Lipids

Lipids are a broad family of molecules including fats, waxes and some vitamins (including vitamins A, D, E and K) that are 'hydrophobic' – they repel water and are only soluble in organic solvents such as acetone. Lipids have a wide range of biological functions, including storing energy, maintaining cell membranes and acting as hormones that are able to coordinate complicated processes like fertility (see page 222).

Common lipid types include fats, steroids and phospholipids. Fats store energy and cushion organs, protecting them from damage, and are composed of fatty acids and glycerol, a sweet-tasting alcohol. Steroids contain four ring-shaped hydrocarbon molecules and include the dietary fat cholesterol as well as the sex hormones estradiol and testosterone.

Phospholipids usually contain two fatty acids and a phosphate group. In water, they arrange themselves into a two-layered sheet with all their hydrophobic tails lined up in the middle. This layered structure forms cell membranes, which regulate the flow of ions and molecules in and out of a cell.

Phospholipids within a cell membrane

1 Individual phospholipid unit
2 Polar heads attracted towards water
3 Watery external environment
4 Non-polar tails face away from water
5 Hydrophobic internal environment repels water

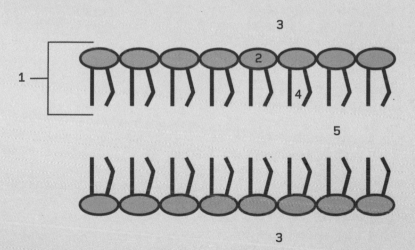

Metabolism

Metabolism describes the host of chemical reactions required to keep living organisms alive, generating energy for essential growth and reproduction – for instance, healing injuries and eliminating toxins.

Apart from water, most molecules in living organisms are amino acids, the building blocks of proteins, as well as carbohydrates and lipids. Metabolic reactions involving them divide into two categories. 'Anabolism' builds molecules like proteins during construction of new cells and tissues, while 'catabolism' breaks down molecules from food for use as a source of energy.

Enzymes also play a vital role in metabolism by acting as catalysts that efficiently transform one chemical into another, helping amino acids join up to form proteins, for instance, or breaking down dietary starch into its component sugars. Healthy metabolism in humans depends on good nutrition, plentiful water and exercise. A lack of any one of these decreases your metabolic rate and can lead to weight gain.

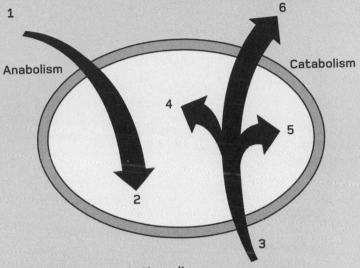

Metabolic pathways in the cell

1 Nutrients for biosynthesis are taken into cell
2 Nutrients used to make new cell components
3 Food is broken down to produce energy

4 Energy used for growth
5 Energy for movement and transport of nutrients
6 Waste products

Chemosynthesis

Chemosynthesis is the process by which some exotic microbes living in hot deep-sea vents derive their energy. It's similar to photosynthesis (see page 176), but doesn't use sunlight. Instead, the energy comes from the oxidation of inorganic chemicals such as hydrogen sulphide bubbling up from the Earth's crust.

In hydrothermal vents, geothermal heat coming up from fissures on the ocean floor can heat water to more than 100°C (212°F). Amazingly, some bacteria called extremophiles thrive in these vents at temperatures up to about 120°C (250°F). There's no sunlight available, so the bacteria produce their energy by turning available chemicals into sugar. For instance, some bacteria oxidize hydrogen sulphide and use the energy stored in its chemical bonds to make glucose from water and carbon dioxide dissolved in sea water.

Scientists speculate that these bacteria would have been well adapted to the hot conditions on the early Earth, making them a good candidate for one of the earliest types of life.

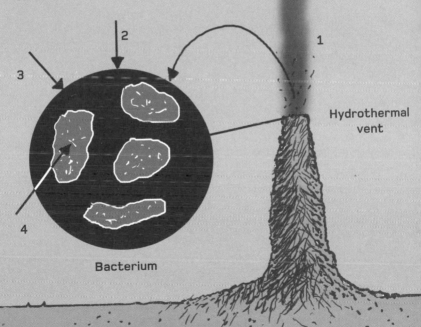

1 Hydrogen sulphide produced by vent
2 Bacterium absorbs dissolved carbon dioxide from seawater
3 Bacterium absorbs water
4 Sugars produced within bacterium

Hydrothermal vent

Bacterium

Receptors

In biochemistry, a receptor is a protein molecule in the membrane or cytoplasm of a cell, onto which signalling molecules such as hormones (see page 224) attach to deliver chemical instructions. For example, the hormone insulin regulates blood sugar by latching onto a receptor in muscle or liver cells, triggering reactions that speed up sugar absorption.

Signalling molecules effectively target specific receptors because they have the right size, shape and electric charge distribution to grab hold of them, a bit like a key fitting into a lock. The binding 'unlocks' the cell to cause chemical changes. Many drugs mimic signalling molecules to promote their effect. For instance, morphine mimics endorphins, naturally occurring feel-good chemicals in the body that relieve pain.

Other drugs lock onto receptors purely in order to block up available binding sites, inhibiting the effects of natural signalling molecules. Examples include the antihistamines, which alleviate allergies by inhibiting chemicals called histamines that cause rashes, sneezing and itching.

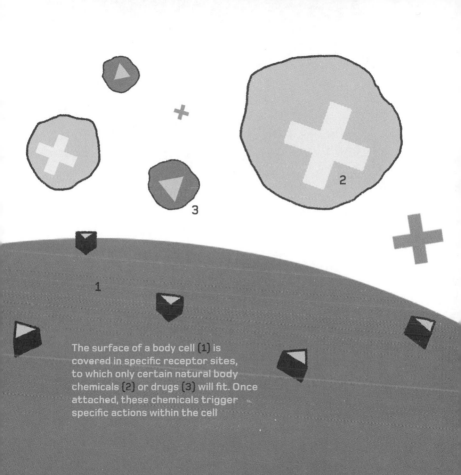

The surface of a body cell (1) is covered in specific receptor sites, to which only certain natural body chemicals (2) or drugs (3) will fit. Once attached, these chemicals trigger specific actions within the cell

DNA

D eoxyribonucleic acid, or DNA, is the molecule that encodes the genetic instructions for the development and function of all living, self-replicating organisms. Nearly every cell in a person's body has the same DNA, which is mostly inside the cell nucleus, although some of it resides in mitochondria (see page 162).

The information in DNA is stored as a sequence of four chemical 'bases' called adenine (A), guanine (G), cytosine (C) and thymine (T). The bases team up (A with T, C with G) to form units called base pairs. Human DNA consists of about 3.2 billion base pairs. Each base is attached to a sugar molecule called deoxyribose and a phosphate molecule to form a 'nucleotide'.

The nucleotides are arranged in two long strands that form a double helix shaped like a spiralling ladder, with the base pairs forming the rungs and the sugar and phosphate molecules forming the vertical supports. DNA replicates itself by splitting into single strands that serve as templates for duplicating the sequence of bases.

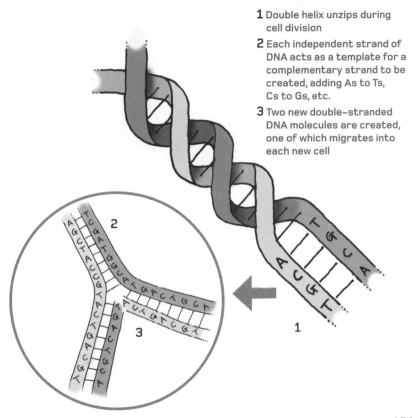

1 Double helix unzips during cell division

2 Each independent strand of DNA acts as a template for a complementary strand to be created, adding As to Ts, Cs to Gs, etc.

3 Two new double-stranded DNA molecules are created, one of which migrates into each new cell

RNA

Ribonucleic acid or RNA is a molecule that has some similarities to DNA but does not store genetic information (except in RNA viruses). Instead, RNA performs many different functions in a cell, acting as a temporary copy of genetic information, for instance.

Like DNA, RNA molecules have a sequence of four chemical 'bases', in this case uracil (U), adenine (A), cytosine (C) and guanine (G). Each base is also attached to a sugar molecule (ribose) and a phosphate molecule to form a 'nucleotide'. The bases sometimes team up with each other (U with A, C with G) to form a double helix structure like DNA, but RNA usually occurs as a single strand.

'Messenger' RNA (mRNA) is a short-lived molecule that copies a cell's DNA and carries it to the cell's protein synthesis machinery, the ribosome, which reads off its information to make the right protein. 'Transfer' RNA (tRNA) molecules latch onto individual amino acids and frogmarch them to the ribosome, where they are incorporated into proteins.

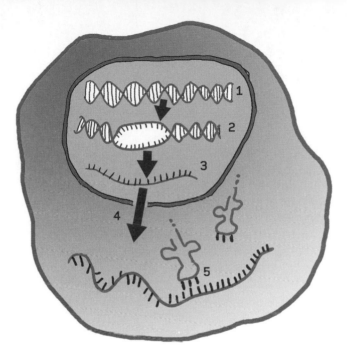

1 DNA
2 Double helix of
   DNA unzips
3 mRNA

4 Messenger RNA is transported
   out of nucleus to ribosome
5 tRNA brings amino acids to build
   new proteins in the ribosome
   (see page 164)

Genes

Genes are segments of DNA that act as a blueprint for the production of specific proteins. They exist in alternative forms called alleles, which determine the distinct traits that parents pass on to their offspring according to the laws of inheritance (see page 156). The entirety of an organism's genetic information is called the genome.

The 13-year Human Genome Project, completed in 2003, was a mammoth international effort to identify the sequences of the 3.2 billion base pairs in human DNA and identify its approximately 25,000 genes. This showed that the average gene consists of about 3,000 base pairs, but sizes vary enormously, with the largest gene having 2.4 million base pairs.

The Human Genome Project has clarified the role that certain gene sequences play in diseases, including breast cancer and muscular dystrophy. However, it also revealed that only about 2 per cent of the genome actually encodes instructions for protein synthesis, and the role of some of the remaining DNA remains a mystery.

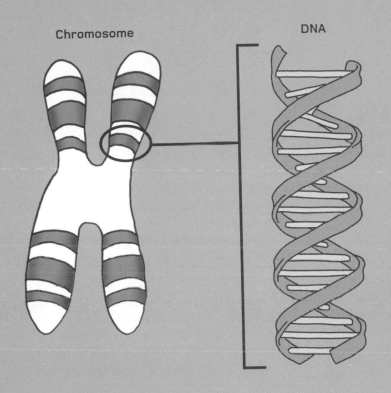

Chromosome

DNA

Laws of inheritance

The laws of inheritance are basic rules about how animals and plants pass on traits to their offspring. Gregor Mendel, a 19th-century Austrian monk, discovered the laws in pea-breeding experiments, in which he cross-fertilized pea plants and studied traits like flower colour and the length of plant stems in subsequent generations.

Mendel's experiments revealed that two factors (now known to be genes) determined the traits, one from each parent, and if the two inherited factors are different, the offspring expresses just one of them, the so-called 'dominant' trait. He also noticed that different traits like flower colour and stem length are inherited independently.

Today we know that all the human blood groups (A, AB, B or O) are determined by a single gene. A and B are dominant while O is 'recessive', so a child who inherits A + O or B + O from its parents will have blood types A and B respectively. A and B are said to be 'co-dominant', so that inheriting both A and B gives blood type AB.

Laws of heredity demonstrated by blood types

		Allele inherited from mother		
		A	B	O
Allele inherited from father	A	A	AB	A
	B	AB	B	B
	O	A	B	O

Blood group of
of offspring

Prokaryotes

Prokaryotes are simple single-celled organisms that don't have a cell nucleus that houses DNA. Instead, their DNA forms a free-floating bundle in the middle of the cell. Like eukaryotic cells (see page 160), prokaryotes have ribosomes where amino acids are assembled into proteins, and sometimes a 'flagellum' – a rudder-like tail that propels them.

The fossil record reveals that prokaryotes evolved on Earth very early, at least 3.5 billion years ago. They reproduce asexually (see page 167) and are typically about 1–10 micrometres (millionths of a metre) wide. They fall into two subcategories: bacteria and archaea.

Bacteria were discovered in the late 1600s and are ubiquitous in every habitat on Earth. They have a wide range of shapes, including spheres, spirals and rods. Archaea were first classified as a separate group in the late 1970s. Most look similar to bacteria but they have a completely different genetic and biochemical make-up, and often inhabit extreme habitats such as scorching hydrothermal vents on the ocean floor.

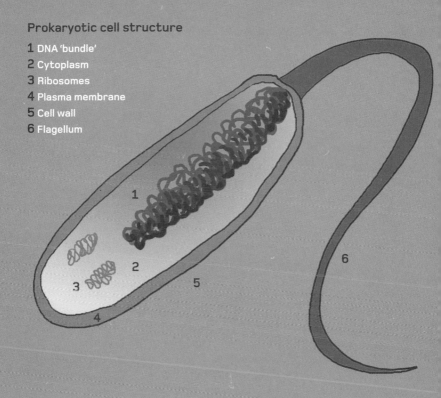

Prokaryotic cell structure

1 DNA 'bundle'
2 Cytoplasm
3 Ribosomes
4 Plasma membrane
5 Cell wall
6 Flagellum

Eukaryotes

A long with prokaryotes (see page 158), eukaryotes are one of two major cell types – they make up everything from single-celled amoebas to complex animals and plants. A typical size for a eukaryotic cell is roughly 0.01 mm across – about 10 or 15 times wider than a typical prokaryote.

The 'plasma membrane' forms the outer barrier of a eukaryotic cell, while the cell nucleus houses DNA packaged into chromosomes that vary widely between organisms. Humans have 23 pairs of large, linear chromosomes.

The nucleus is surrounded by a water-rich fluid called cytosol and various organelles that perform different tasks. Mitochondria (see page 162) generate energy, while the endoplasmic reticulum is a network of interconnected membranes dotted with ribosomes, which assemble proteins.

Fossil evidence suggests eukaryotes evolved at least 1.7 billion years ago. One possibility is that they arose because some prokaryotic cells engulfed others, which were not digested but lived on as organelles and reproduced.

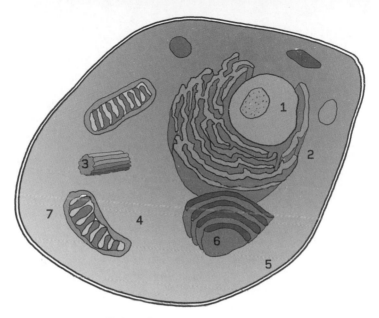

Eukaryotic cell structure

1 Nucleus containing DNA

2 Endoplasmic reticulum carrying ribosomes

3 Centriole plays a key role in cell division

4 Cytosol (intracellular fluid)

5 Plasma membrane

6 Golgi apparatus sorts and modifies proteins

7 Mitochondrion

Mitochondria

Mitochondria act as the power plants of eukaryotic cells (see page 160), converting the energy from food into a form that cells can use. A cell may have hundreds or even thousands of mitochondria depending on its energy demands.

Mitochondria are like little factories that specialize in using energy from reactions between oxygen and simple sugars to produce molecules of adenosine triphosphate (ATP), the cell's main energy source. ATP is a bit like a charged battery — removal of a phosphate group releases energy to drive complex reactions, leaving 'uncharged' adenosine diphosphate (ADP), which returns to mitochondria to be recharged into ATP.

Surrounded by two membranes, mitochondria have their own genetic material and reproduce independently of their host cell. Scientists suspect that the ancient ancestors of mitochondria were probably free-living bacteria that somehow became engulfed in other cells. The bacteria thrived in the protective environment of their new host cells, while the hosts came to rely on the bacteria for energy production.

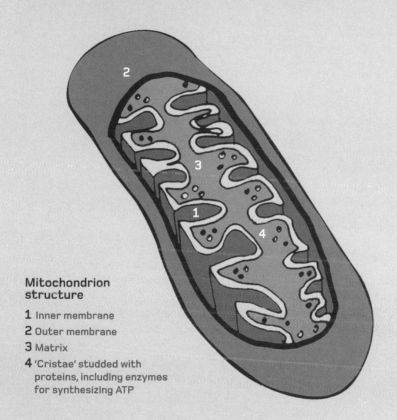

Mitochondrion structure

1 Inner membrane
2 Outer membrane
3 Matrix
4 'Cristae' studded with proteins, including enzymes for synthesizing ATP

Ribosomes

In all cells, including those of plants, animals and bacteria, ribosomes are the mini factories that assemble proteins. Each ribosome is built from RNA molecules and proteins. Some of them are free to roam in the watery cytoplasm of a cell, while others are bound to the endoplasmic reticulum, a complicated network of interconnected membranes inside eukaryotic cells.

A strand of messenger RNA (mRNA) brings a copy of genetic information from a cell's DNA to a ribosome. Meanwhile, transfer RNA (tRNA) molecules latch onto single amino acids and deliver them to the ribosome, to be incorporated into proteins according to the mRNA's instructions.

Cells typically have several thousand ribosomes, but the number can reach several million. The chemical structure of ribosomes is different in bacteria and animal cells, and the differences allow many antibiotic drugs to selectively disrupt the ribosomes of disease-causing bacteria, sabotaging their protein production without making people or animals sick.

Peptide synthesis

1 Incoming tRNA bound to amino acid at one end carries genetic 'key' of chemical bases at its other end

2 Genetic code for constructing peptide chain is held on mRNA

3 'Key' on tRNA matches up with 'lock' on on mRNA strand.

4 Amino acids link together to form growing peptide chain

5 'Empty' tRNA units disconnect from mRNA strand

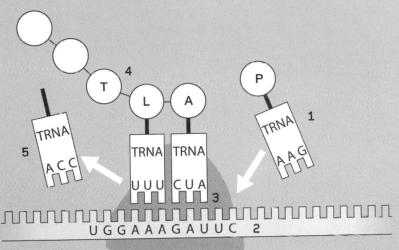

Ribosome

Cell division

Cell division is the process by which biological cells multiply. Eukaryotic cells (see page 160), including human cells, create identical copies of themselves for growth or repair of tissues through a process called mitosis. The double-stranded DNA in the cell nucleus unzips into two strands that each join with nucleotides (see page 150) to form two copies of the original DNA. In the next step, 'cytokinesis', the cell splits into two identical copies of the original cell.

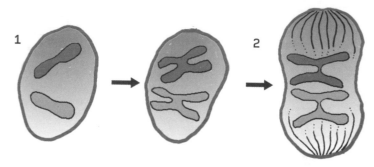

'Meiosis' is a type of cell division that creates eggs and sperm for sexual reproduction. The eggs and sperm cells have half the normal number of chromosomes. When sperm fertilizes an egg, they fuse to regain the normal chromosome quota, half from the male and half from the female.

Prokaryotic cells such as bacteria (see page 158) usually divide in a process called binary fission. The single DNA bundle in a prokaryotic cell replicates, then the two copies stick to different parts of the cell membrane, going their separate ways when the cell splits in two.

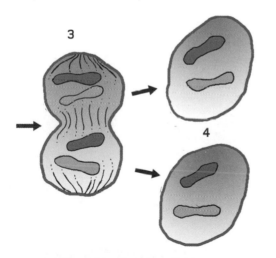

3

4

1 DNA in nucleus unzips into two separate strands
2 Mitosis: DNA strands replicate
3 Cytokinesis: cell splits in two
4 Two new eukaryotic cells

Gametes

Gametes are the germ cells of animals and plants that reproduce sexually. In most animals, including humans, the male gametes are called sperm cells and the female ones are called eggs.

Gametes form in a cell division process called meiosis, in which a cell splits into four copies instead of the usual two, so that the gametes contain half the normal number of chromosomes – just one of each of the 23 pairs normally found in human cells. 'Chromosomal crossover' occurs during meiosis, each chromosome pair exchanging genetic material so that the gametes end up with new combinations of genes.

During fertilization, an egg and sperm fuse to create a 'zygote' with two copies of each chromosome, one from the male and one from the female. This single cell develops into an embryo by cell division. In most mammals, the X and Y sex chromosomes determine gender. Inheriting an X chromosome from both parents gives female offspring (XX), while inheriting X from the mother and Y from the father gives males (XY).

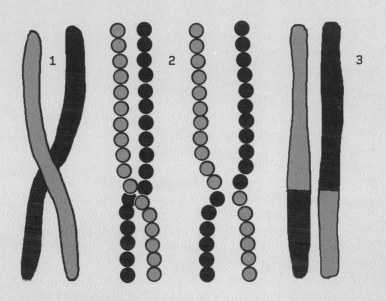

1 One chromosome from each parent
2 Chromosomes exchange genetic material during meiosis
3 Inherited chromosomes have new combinations of genes

Biological classification

The basic classification system for animals, plants and microbes was first introduced in the early 1700s by the Swedish botanist and zoologist Carl Linnaeus, who grouped species according to shared physical characteristics. The groupings have been revised since then to take account of new information about evolutionary trees.

Today many scientists divide all life into three domains: Archaea and Bacteria (see page 158) and Eukaryota, which includes complex animals and plants. The domains are subdivided into kingdoms, usually six: Animalia, Plantae, Fungi, Protista, Archaea, Bacteria. Subcategories of the kingdoms multiply into phyla, classes, orders, families and genera with the final, most specific category being species.

Typically, a species is a group of organisms that are biologically similar enough to each other that they can interbreed to create fertile offspring. The number of species of life on Earth is impossible to measure – scientists guess that it lies anywhere between about 5 million and 100 million.

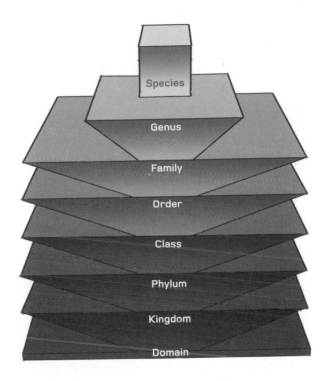

Animals

Animals are multicellular, eukaryotic organisms from the kingdom Animalia, and they form a vast group of more than 1.5 million known species including worms, insects, sponges and people. All animals are 'heterotrophs' – unable to generate some vital organic chemicals internally, they have to eat other organisms to stay alive. As they develop, their body plans become fixed, although some animals undergo metamorphosis – for instance, when a caterpillar pupates and transforms into a butterfly.

Some animals reproduce asexually, including aphids, which sometimes reproduce independently by effectively cloning themselves (see page 210). But the vast majority of animals reproduce by sexual reproduction, in which a male and female combine their genetic material to create their offspring.

Animals usually sexually reproduce when an egg and sperm fuse during fertilization to create a 'zygote' with two copies of each chromosome, one from the female and one from the male. This zygote develops into an embryo by cell division.

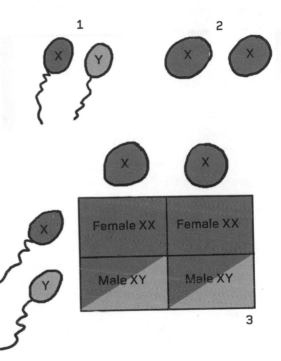

1 Male sperm carry either X or Y sex chromosomes

2 Female eggs always carry X chromosomes

3 Combination of sex cells gives rise to offspring with different genders

Plants

Plants are multicellular eukaryotic organisms that use sunlight to generate energy and organic chemicals through photosynthesis (see page 176). They include familiar organisms like grasses, bushes and trees as well as green algae, which are largely aquatic and exist as single cells, colonies and seaweeds. There are roughly 350,000 known species of plants.

Plants typically have a main stem growing up from a root system in the soil, with branches emerging from 'nodes' on the stem. Sexual reproduction in plants often involves male pollen grains fertilizing the ovules in a flower. Protective seed coats form around the fertilized ovules before they disperse, allowing a new generation of plants to germinate. Asexual reproduction doesn't involve flowers – plants can create genetically identical copies of themselves when bulbs divide, for instance.

The first land plants evolved more than 450 million years ago, while forests spread on land around 385 million years ago. Flowering plants evolved roughly 140 million years ago, and since then have become the dominant land plants.

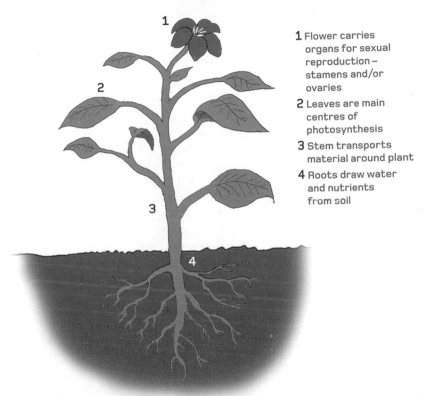

1 Flower carries organs for sexual reproduction – stamens and/or ovaries

2 Leaves are main centres of photosynthesis

3 Stem transports material around plant

4 Roots draw water and nutrients from soil

Photosynthesis

Photosynthesis is the process in which plants, as well as some bacteria and eukaryotic microorganisms, use the energy from sunlight to produce sugars such as glucose. Plant leaves are like solar energy collectors crammed full of photosynthetic cells. These cells combine water and carbon dioxide molecules to create sugars and oxygen.

Water enters a land plant through its roots and is then transported up to the leaves. Atmospheric carbon dioxide enters the leaves through tiny pores called stomata, which open and close depending on environmental conditions. Likewise, oxygen produced during photosynthesis passes back out into the atmosphere through the stomata.

During respiration, plants combine sugar with oxygen to produce carbon dioxide and water, in the process creating adenosine triphosphate, or ATP (see page 162), the molecule that supplies energy for essential work like protein building. Plant respiration dominates at night, when the photosynthetic uptake of carbon dioxide and release of oxygen stops.

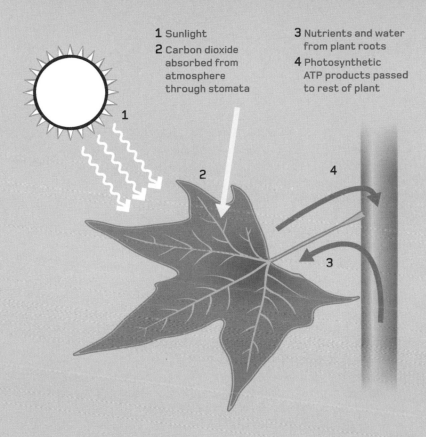

1 Sunlight
2 Carbon dioxide absorbed from atmosphere through stomata
3 Nutrients and water from plant roots
4 Photosynthetic ATP products passed to rest of plant

Prokaryotic microbes

Microbes made of prokaryotic cells (see page 158) fall into two classes: bacteria and archaea, which are both single-celled. These microbes are the most diverse and abundant group of organisms on Earth, accounting for at least half of all biomass, despite their tiny size (they are typically about one-thousandth of a millimetre wide).

Bacterial cells come in a wide variety of shapes. Spherical ones are called cocci, while elongated rod-shaped ones are called bacilli. Sometimes, bacteria come in pairs, where their name is prefixed with 'diplo-'. Those that form long chains are prefixed 'strepto-', while some bacteria form triangular groups prefixed 'staphylo-'. Rod-shaped bacteria can divide to form a picket-fence structure called a palisade arrangement.

Many bacteria cause diseases. Streptococcus species can cause pneumonia and meningitis, for instance. Archaea look similar to bacteria, but have a completely different biochemical make-up and do not have a known role in disease. Archaea were possibly the earliest life forms on Earth.

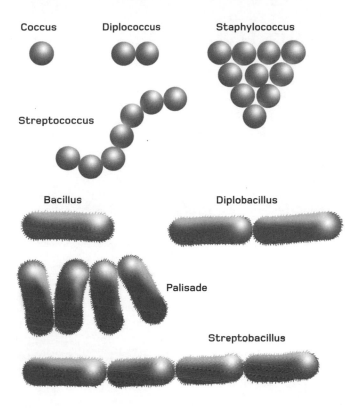

Coccus   Diplococcus   Staphylococcus

Streptococcus

Bacillus   Diplobacillus

Palisade

Streptobacillus

Eukaryotic microbes

Eukaryotic microbes are a diverse range of organisms so small that they are invisible to the naked eye. They divide into four groups: animals, plants, fungi and 'protists'. Unlike the simpler prokaryotic microbes, the bacteria and archaea, eukaryotic microbes house their DNA inside a cell nucleus.

Micro-animals include dust mites and many nematodes (roundworms) as well as rotifers, tiny filter feeders usually found in fresh water. Some microscopic, photosynthetic green algae are classed as plants. Fungi have several single-cell species, including baker's yeast. Protists are a diverse group of organisms that have little in common except simplicity – they are single-celled or multicellular without specialized tissues.

Like bacteria, eukaryotic microbes can cause serious diseases, including malaria, while some fungi present a major health hazard to crops. Finding treatments for these diseases is challenging because any chemical that kills eukaryotic organisms or inhibits their growth is also likely to be toxic to the eukaryotic cells of plants or animals.

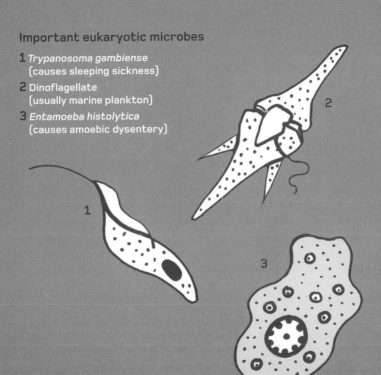

Important eukaryotic microbes

1 *Trypanosoma gambiense*
  (causes sleeping sickness)
2 Dinoflagellate
  (usually marine plankton)
3 *Entamoeba histolytica*
  (causes amoebic dysentery)

Viruses

Viruses are tiny packets of genetic material capable of infecting the cells of living organisms, including animals, plants and bacteria. Viruses can't reproduce on their own. Instead, they breed by invading the cells of a host and hijacking their replication machinery.

Viruses consist of RNA or DNA protected by a protein coating, and are typically only 10–300 nanometres (billionths of a metre) in size. They infect organisms by penetrating cell membranes, releasing their genetic cargo and forcing the doomed cells to replicate them. New viruses assemble inside each host cell before bursting out, killing the cell. However, some viruses can remain dormant inside cells for years.

Plant viruses are often transmitted between plants by insects that feed on them. Human colds and flu spread through coughing and sneezing, while several viruses are transmitted by sexual contact, including the human immunodeficiency virus (HIV). Fortunately, our immune systems successfully combat most viral infections, while vaccinations can prevent others.

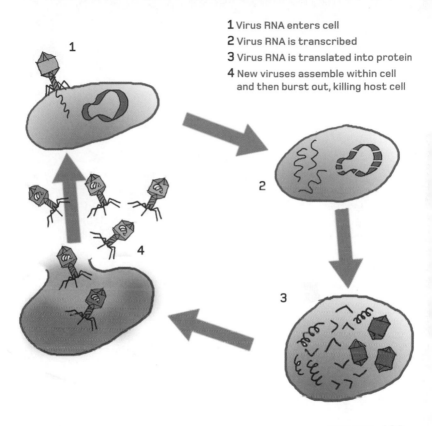

1 Virus RNA enters cell
2 Virus RNA is transcribed
3 Virus RNA is translated into protein
4 New viruses assemble within cell and then burst out, killing host cell

Origins of biochemicals

Scientists can only speculate on how the organic compounds necessary for life emerged on Earth. They could have arisen spontaneously from chemical reactions of simpler compounds, or they could have arrived from space.

A famous experiment at the University of Chicago in 1953 tested whether stormy conditions on the young planet could have triggered reactions between simple chemicals to create the ingredients of life. Stanley Miller and Harold Urey zapped a mixture of water, methane, hydrogen and ammonia with electrical discharges to simulate lightning. Sure enough, this cooked up many organic compounds including amino acids, the building blocks of proteins.

Since then, evidence has emerged that vigorous volcanism on the early Earth enriched the chemical soup with carbon dioxide, nitrogen and sulphur compounds. These could have spurred the production of biochemicals. And astronomers have identified organic molecules including amino acids in comets, so comet impacts could have delivered off-the-shelf biochemicals.

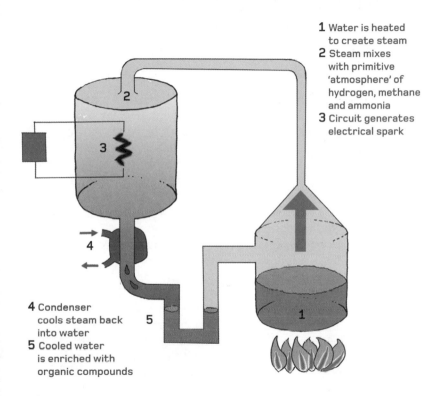

1 Water is heated to create steam
2 Steam mixes with primitive 'atmosphere' of hydrogen, methane and ammonia
3 Circuit generates electrical spark
4 Condenser cools steam back into water
5 Cooled water is enriched with organic compounds

Replicating life

The early Earth somehow acquired the complex organic chemicals necessary for life. But how did the jump from inanimate chemicals to living, self-replicating organisms occur?

One puzzle about life's origins is that all self-replicating organisms today use DNA to store the genetic information needed to build proteins, while protein enzymes are necessary to copy the DNA and spawn new generations. In other words, DNA and proteins are both vital to life, but the chances that the Earth cooked them up simultaneously seem slim.

An alternative idea is the 'RNA world' hypothesis, which proposes that the earliest replicating organisms depended solely on RNA (see page 152), which is better at multitasking than DNA – it can act as an enzyme as well as a carrier of genetic codes. However, many scientists are unconvinced that complex RNA molecules could have spontaneously assembled in early-Earth conditions. Ultimately, it may be impossible to find a convincing theory for how life arose unless a simple 'one-pot' experiment repeats it in a very compelling way.

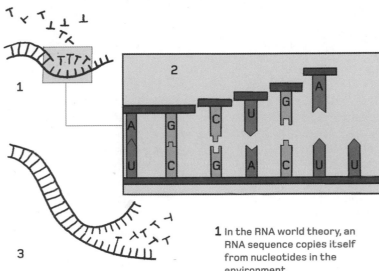

1 In the RNA world theory, an RNA sequence copies itself from nucleotides in the environment

2 Each base can match with only one other type of partner base

3 New sequence unzips and the process starts again

Extraterrestrial origins of life

There's one intriguing alternative solution to the mystery of how life arose on Earth – it didn't. A theory called panspermia suggests that comets and meteors from space not only delivered complex organic chemicals to the young Earth, they also delivered bona fide living organisms from which all life evolved. In that case, we are all descended from aliens.

Proponents of panspermia suggest that simple life is widespread on bodies like comets throughout the solar system, and possibly beyond. Comets that hit Earth, delivering much of the water in its oceans, could also have brought fully functional living microbes that had gradually evolved in space over billions of years. That would explain why life 'emerged' on Earth so astonishingly quickly after our planet became inhabitable.

What's more, there is intriguing evidence that some hardy bacteria can survive the harsh conditions of space, and even a violent impact on a planetary surface. But panspermia remains very speculative – although there are hints that bugs could exist in space, there is no direct evidence that they actually do.

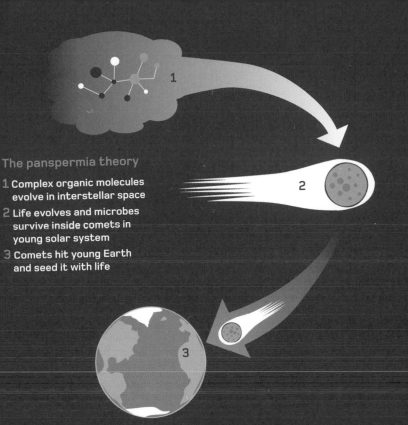

The panspermia theory

1 Complex organic molecules
  evolve in interstellar space
2 Life evolves and microbes
  survive inside comets in
  young solar system
3 Comets hit young Earth
  and seed it with life

Evolution

Evolution describes the way populations of living organisms change over time due to changes in heritable genetic traits, such as eye colour in humans. Sometimes the changes are due to environmental pressures – for instance, giraffes may have evolved long necks because those eating plentiful foliage high in the trees survived to produce most offspring.

This 'natural selection' (see page 192) is one key driver of evolution, but other genetic factors come into play.

Spontaneous, random mutations of genes can sometimes create a beneficial trait that helps an organism produce more offspring, so the mutation persists in the population. 'Genetic drift' (see page 194) also plays a role – certain gene variants might flourish simply by chance.

Sometimes two or more species can evolve through co-evolution, when they have close ecological interactions with each other. For instance, a plant might evolve thorns to deter herbivores, while the herbivores in turn evolve defences against thorns that thwart the plant's strategy.

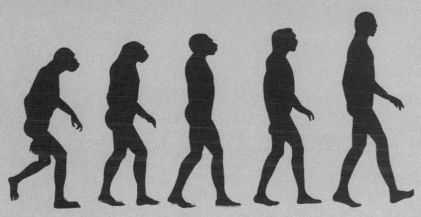

Natural selection

Natural selection is one of the basic mechanisms of evolution. The British scientists Charles Darwin and Alfred Russel Wallace independently proposed the theory in 1858.

Organisms have different traits, such as size, and these influence whether or not an organism survives long enough to pass on their traits to their offspring. Advantageous traits become more common in successive generations. Over time, populations split into different species ('speciation'), and looking back in time, any pair of organisms shares a common ancestor. For instance, humans shared a common ancestor with chimpanzees about 6 million years ago.

Peppered moths offer an example of rapid natural selection during the industrialization of the United Kingdom. Common pale-coloured peppered moths became conspicuous on tree trunks that had darkened with soot, making them easy prey for birds, while darker ones preferentially survived long enough to breed. Eventually, the population became primarily dark, but the process reversed when the clean air standards were enforced.

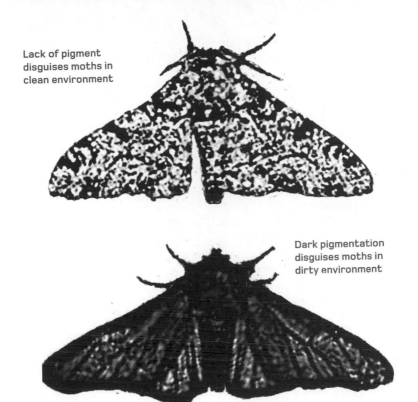

Lack of pigment disguises moths in clean environment

Dark pigmentation disguises moths in dirty environment

Genetic drift

Genetic drift is one of the driving forces behind evolution. It describes the way that certain traits in a population might flourish or vanish by chance, because individuals with these traits happen to breed most offspring or not breed at all.

Genetic drift tends to rapidly reduce genetic diversity in very small populations. For instance, if only two individuals in a population of ten animals carry a certain gene variant and do not breed fertile offspring, that variant will be gone from the population for good.

A special case of genetic drift is the 'founder effect', which occurs when a small number of individuals become isolated from the main population. For example, in the late 1700s, a typhoon on the island of Pingelap in Micronesia left only around 20 human survivors to create future generations. Today, 5–10 per cent of the island's population suffer from a total colour blindness disorder that is extremely rare elsewhere, because one of the typhoon survivors must have carried a recessive gene linked to the disorder.

Rapid genetic drift over three generations

1 First generation – frequency of 'rare' trait 17 per cent

2 Second generation – frequency of rare trait increases to 25 per cent

3 Third generation – frequency of rare trait rises to 39 per cent

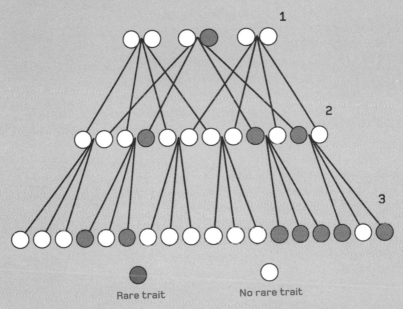

Rare trait   No rare trait

Human origins

All people living today are descended from a single female dubbed 'mitochondrial Eve' who lived in Africa about 110,000–130,000 years ago. Scientists deduced this from modern genetic analysis of the DNA in human mitochondria (see page 162), which is inherited via the maternal line.

The widely accepted 'Out of Africa' theory suggests that modern humans (*Homo sapiens*) first evolved in Africa roughly 200,000 years ago, then migrated out all over the world during the last 100,000 years. They reached the Middle East by about 70,000 years ago, south Asia by 60,000 years ago and western Europe around 40,000 years ago. The dates for North American colonization are unclear – this may have started around 30,000 years ago or much later.

These human settlers largely replaced indigenous pre-human species, such as the heavy-browed *Homo erectus* in Asia. However, genetic evidence hints that some interbreeding took place between anatomically modern humans and Neanderthals, members of the genus *Homo* that died out 30,000 years ago.

The 'Out of Africa' theory

1 Human origins in eastern Africa
(around 200,000 years ago)

2 Expansion across Africa
(100,000 years ago)

3 Migration into Asia
(60,000 years ago)

4 Expansion into southeast
Asia and Australia
(50,000–60,000 years ago)

5 Migration into Europe
(40,000 years ago)

6 Colonization of Americas
(15,000–35,000 years ago)

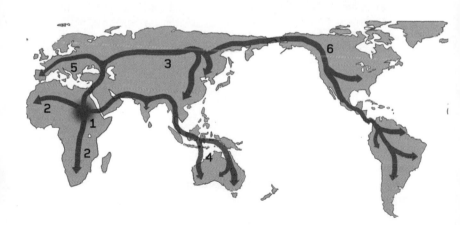

Food webs

Food webs and food chains are graphical representations of the feeding relationships between living organisms in an ecosystem. A food chain usually describes a linear sequence such as 'hawks eat snakes, snakes eat toads', while a food web maps a more complicated network of interconnected chains.

Living organisms can be divided into three categories: producers, consumers and decomposers. Producers such as green plants can make their own food from energy and simple inorganic compounds. Consumers eat other organisms. They include herbivores ('primary consumers') that eat plants, carnivores that eat animals and omnivores that eat both. Animals that eat primary consumers (such as snakes that eat mice) are called secondary consumers, while tertiary consumers eat secondary ones, and so on.

Decomposers are organisms that feed on dead animals and plants, such as earthworms or fungus growing on a dead log. They break down materials into the simple inorganic chemicals and nutrients vital for the producers, such as green plants.

1 Sunlight
2 Producer
3 Primary consumer
4 Secondary consumer
5 Tertiary consumer
6 Decomposers

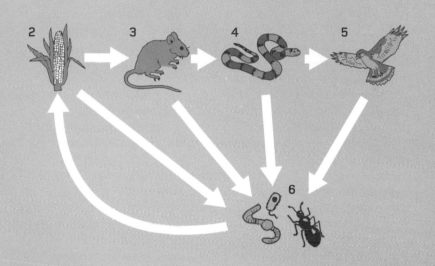

Cycles

The carbon cycle describes the way that carbon, a crucial ingredient for all organic substances, circulates through different components of the Earth's environment, including the land, oceans, atmosphere and planetary interior.

Plants absorb carbon dioxide (CO_2) during photosynthesis, and release carbon as they die and decompose. Burial and compression over millions of years can transform them into fossil fuels. During respiration, plants and animals release CO_2, and burning fossil fuels produces more. The gas is slightly soluble in water, so lakes and oceans absorb some of it, and organisms such as corals and shellfish convert it into calcium carbonate that accumulates in ocean sediments when they die.

Other important cycles include the water cycle, in which warm air passing over the sea surface evaporates water, which rises and condenses into clouds before falling as precipitation. In the nitrogen cycle, nitrogen 'fixing' locks up nitrogen in plants as nutrients, while soil bacteria eventually break nitrogen compounds down, returning nitrogen gas to the atmosphere.

Major elements of the carbon cycle

1 Carbon in atmosphere

2 Carbon absorbed by
living organisms

3 Carbon released by
living organisms

4 Carbon cycled to and
from oceans

5 Carbon buried in rocks

6 Carbon released by
fossil fuel

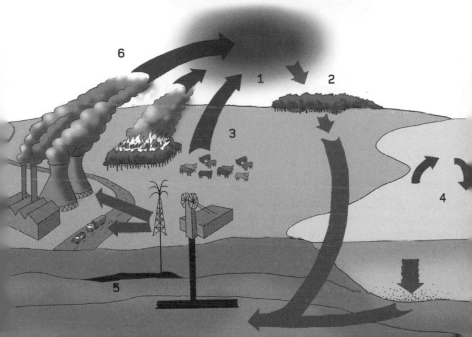

Biodiversity

Biodiversity measures how much variety there is between all the different species of life on Earth, from single-celled bacteria through insects to blue whales, the largest known animal ever to have existed. The term also sometimes describes the genetic diversity within a single species, or the diversity of ecosystems like wetlands and forests.

Around 1.75 million species of living organisms have been

Scientists rate the conservation status of a species on a scale from extinct to least concern – the three middle categories are all classed as 'threatened'.

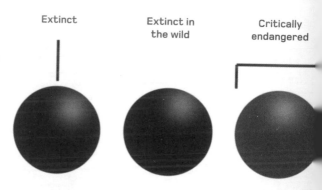

Extinct

Extinct in the wild

Critically endangered

identified on Earth so far, mostly small ones like bacteria and insects, and estimates suggest the true number could be as high as 100 million. But in recent centuries, there has been a rapid increase in the rate of species extinctions due to human activities such as habitat destruction for farming.

Between 1500 and 2009, international organizations documented more than 800 species becoming extinct, including the Javan tiger that died out completely in the 1980s, but the vast majority of disappearances probably go unnoticed. Conservationists grade the vulnerability of species according to a scale that runs from 'extinct' to 'least concern'.

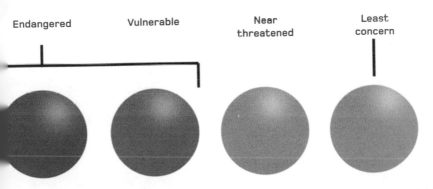

Endangered   Vulnerable   Near threatened   Least concern

Mass extinctions

Mass extinctions occur when environmental conditions change and kill off vast numbers of species of life. A classic example is the Cretaceous–Tertiary extinction, which wiped out the Earth's dinosaurs about 65 million years ago. The consensus is that a large asteroid hit the Earth and altered its climate by filling the atmosphere with dust, blocking sunlight and making the climate too cool for dinosaurs to survive.

The fossil record suggests that many other mass extinctions have occurred in the past, including the Late Ordovician extinction 440 to 450 million years ago. Scientists suspect that this occurred when inhospitable ice sheets grew across an ancient supercontinent called Gondwana, which existed long before the current familiar continents took shape.

Many scientists have speculated that humans are causing a current mass extinction today, by hunting, generating pollution and destroying habitats, particularly 'biodiversity hotspots' such as tropical forests. Many species living in these areas will probably become extinct before we even know they existed.

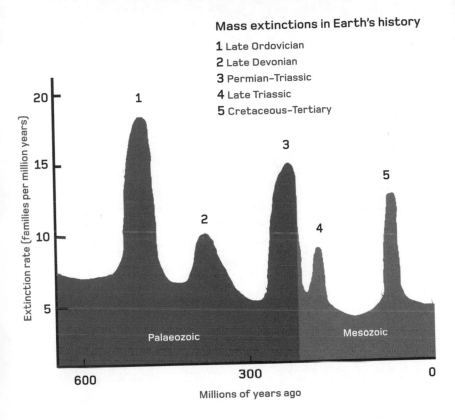

Genetic modification

Genetic modification is the use of modern biotechnology techniques to change the genes of an organism, altering the DNA that instructs its cells how to build proteins. Many crop plants are genetically engineered to possess desirable traits such as resistance to pests or harsh environments.

In traditional breeding of crops and livestock, farmers pick plants or animals with desirable traits and crossbreed them

1 Bacterial cell with pest-killing gene

2 Enzyme extracts gene

3 DNA inserted into plant cell

4 Cell cultured

5 Pest-resistant plant grown

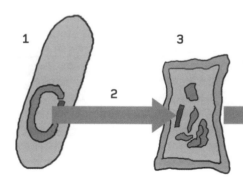

to create commercially valuable offspring. Genetic modification allows the traits of organisms to be altered in ways that are not possible through traditional breeding.

For example, some cotton plants are modified to carry a gene from soil bacteria. This makes them produce a chemical that kills insect pests, reducing the need for pesticides. Sometimes, genetic modification turns down or 'silences' the activity of genes that an organism already has. This can prevent oilseed rape crops producing unhealthy oils, for instance. Genetically modified animals are often used in experiments to study gene functions, but are not yet bred for commercial agriculture.

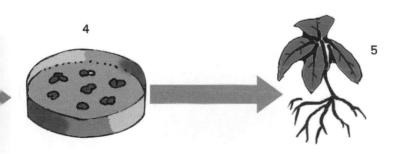

4

5

Pharming

In biotechnology, pharming (linking 'pharmaceutical' and 'farming') means genetically engineering plants or animals so that they produce useful drugs or industrial chemicals. For instance, plants could be genetically engineered so that their seeds contain rich sources of human antibodies, specialized proteins of the immune system that could fight diseases such as cancer, hepatitis and malaria.

1 Eggs extracted and fertilized in vitro

2 Genetic code of fertilized eggs modified in laboratory

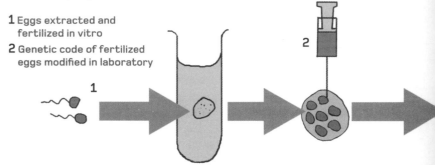

Goats have already been genetically modified for commercial production of a drug called ATryn, which prevents dangerous blood clots. The drug's active protein is purified from the goats' milk. Many plants that have been genetically modified to produce drugs are undergoing tests, but plant-pharmed drugs are not commercially available so far. While supporters say this is a safe and cheap way to manufacture vital vaccines, opponents worry about the plants crossbreeding with natural plants and contaminating the environment and food supplies.

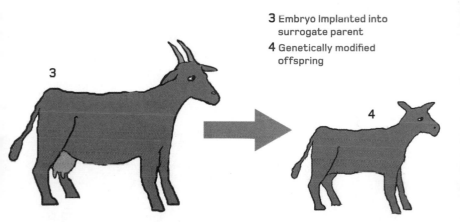

3 Embryo Implanted into surrogate parent
4 Genetically modified offspring

3

4

Cloning

Reproductive cloning is the creation of a genetically identical copy of an organism, with exactly the same DNA. Scientists created the first cloned mammal, Dolly the sheep, at the Roslin Institute near Edinburgh in Scotland in 1996.

Dolly was created by a technique called somatic cell nuclear transfer. Scientists took a cell from an adult sheep's mammary gland and transplanted its DNA-containing nucleus into an egg cell from which the nucleus had been removed. The cell developed into a normal embryo, which was implanted into a surrogate mother sheep who carried the fetus to term and gave birth to Dolly, an exact genetic replica of the adult female that donated the cell nucleus.

Since Dolly, researchers have cloned many large and small mammals including horses, goats, cows, mice, pigs, cats and rabbits. They hope that 'therapeutic cloning' (a type of stem cell therapy) may one day provide genetically matched tissues and organs that can be used in transplant operations without any risk of the patient rejecting them (see page 252).

1 Body cell taken from male
2 Egg cell extracted from female
3 Nucleus removed from egg
4 DNA from male fused with egg cell from female
5 Embryo implanted in surrogate mother
6 Offspring is clone of original male

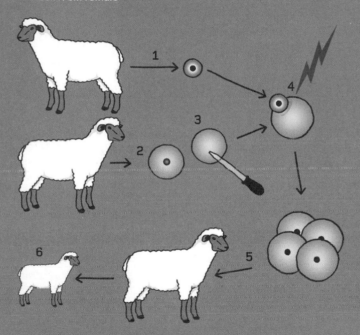

Cardiovascular system

The cardiovascular system circulates blood around the body, transporting oxygen from the lungs and nutrients to organs, muscles and nerves. The heart pumps oxygen-rich blood through a network of blood vessels, the arteries. When it reaches tiny vessels called capillaries in tissues, blood releases the oxygen, which cells use to make energy.

The cells also release waste products, such as carbon dioxide, which the blood absorbs and carries away. The used or 'deoxygenated' blood travels along veins and back to the lungs, where it absorbs fresh oxygen and begins the cycle again. At rest, a normal heart typically beats around 70 to 80 times a minute as electrical impulses make the cardiac muscles rhythmically contract.

Each side of the heart is divided into an upper chamber called an atrium and a larger, lower chamber, called a ventricle. The atria are the blood-receiving chambers and the ventricles are the discharging chambers. Blood flows from each atrium through a one-way valve into the ventricle below.

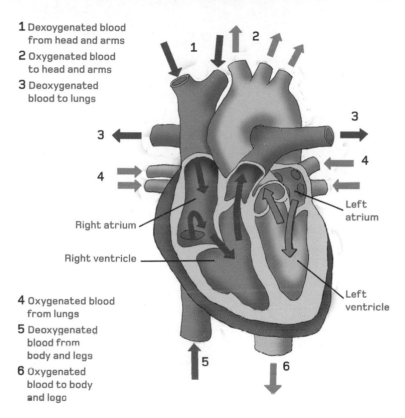

1 Dexoygenated blood
  from head and arms
2 Oxygenated blood
  to head and arms
3 Deoxygenated
  blood to lungs

3

3

4

4

Left
atrium

Right atrium

Right ventricle

Left
ventricle

4 Oxygenated blood
  from lungs
5 Deoxygenated
  blood from
  body and legs
6 Oxygenated
  blood to body
  and logo

5

6

Respiratory system

The respiratory system supplies the blood with vital oxygen for all the body's organs. We breathe in when a sheet of muscles across the bottom of the chest cavity (the diaphragm) contracts, pulling air into the lungs, and we breathe out when the diaphragm relaxes.

Air enters through the mouth and the nose, and passes through the larynx to the trachea or windpipe. This tube splits into two smaller ones called bronchi inside the chest cavity, then divides repeatedly in the lungs, connecting to millions of tiny air-filled sacs called alveoli, surrounded by tiny capillaries. Oxygen diffuses into arterial blood through the capillary walls. Meanwhile, veins fill the alveoli with carbon dioxide waste, which flows out through the same airway when we exhale.

Inhaled air is mostly nitrogen (about 78 per cent) and about 21 per cent oxygen. Exhaled air is roughly 78 per cent nitrogen, 16 per cent oxygen and 4 per cent carbon dioxide. So there is a net absorption of oxygen in the body and a net release of waste carbon dioxide.

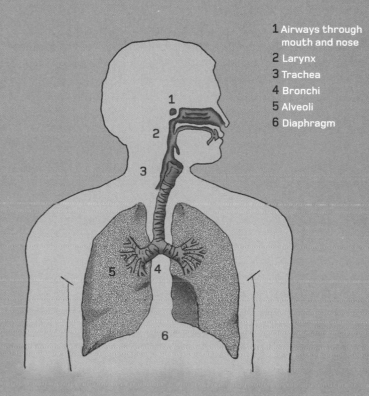

1 Airways through mouth and nose
2 Larynx
3 Trachea
4 Bronchi
5 Alveoli
6 Diaphragm

Gastrointestinal system

The gastrointestinal system is a chain of organs that digest food so the body can absorb nutrients. Swallowed food passes down the oesophagus, which secretes mucus to help food pass easily. Then it enters the stomach, a J-shaped bag.

Glands in the stomach lining secrete juices rich in acid and digestive enzymes, which kill some harmful bacteria and start to break food down. After this primary digestion, food moves into the small intestine. Here the duodenum neutralizes acidity and starts further digestion, which continues in the jejunum and ileum, coiled tubes about 4–6 m (13–20 ft) long in total. When digestive products reach the large intestine through the caecum, almost all nutritionally useful products have been removed. The large intestine removes water from the remains in the colon, before waste is expelled through the anus.

The liver has many important functions, detoxifying substances like alcohol in the bloodstream and producing fat-digesting bile, which is stored in the gall bladder. The pancreas secretes enzymes that aid digestion, as well as hormones.

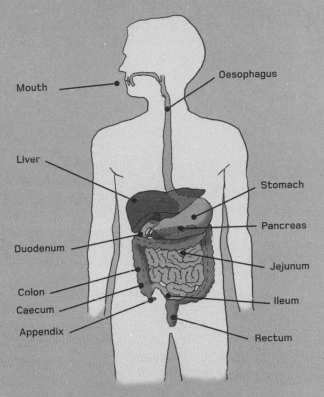

Mouth

Oesophagus

Liver

Stomach

Pancreas

Duodenum

Jejunum

Colon

Caecum

Ileum

Appendix

Rectum

Musculoskeletal system

The musculoskeletal system consists of all the bones in the skeleton along with the muscles, tendons and other connective tissues that support them and allow us to move.

More than 200 bones in humans form a rigid, protective framework to which softer tissues and organs are attached. The skull protects the brain from damage, for instance, while the sternum and rib cage protect the heart and lungs. Bones are connected to each other by bands of fibrous tissue called ligaments. Some bones, including the femur, contain bone marrow that houses stem cells, which can transform into blood cells in order to top up the body's blood supply.

Skeletal or 'voluntary' muscles are bundles of fibres that contract and relax to move bones and joints. They are mostly attached to bones by collagen fibres called tendons and move in response to voluntary instructions from the brain. Smooth or 'involuntary' muscle forms within the walls of organs such as the stomach and intestines, where they push food through the digestive system without any conscious control.

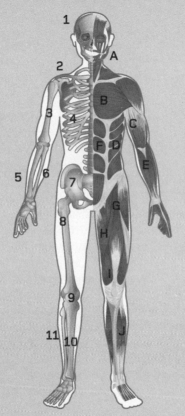

Major bones

1 Cranium
2 Clavicle
3 Humerus
4 Ribs
5 Radius
6 Ulna
7 Pelvis
8 Femur
9 Patella
10 Tibia
11 Fibula

Major muscles

A Sternocleidomastoid
B Pectoralis
C Biceps
D External oblique
E Brachioradialis
F Rectus abdominis
G Rectus femoris
H Sartorius
I Quadriceps
J Tibialis

Urinary system

The urinary system removes the waste products of food digestion from the blood. The waste includes urea, with the chemical formula $(NH_2)_2CO$, which forms when proteins in food break down. The main organs of the urinary system are the kidneys, which also regulate blood pressure, stabilize salt levels and produce a hormone called erythropoietin, which controls the production of red blood cells inside bone marrow.

The kidneys are a pair of purplish-brown organs just below the ribs towards the middle of the back. They remove urea from the blood through tiny filtering units called nephrons, which are balls of small blood capillaries with a small tube or 'renal tubule' attached.

In the nephrons, urea, water and other waste substances form urine that flows through tubes called ureters to the bladder, where urine is stored before excretion through the urethra. Normal urine is sterile – it contains fluids, salts and waste products, but no bacteria or viruses.

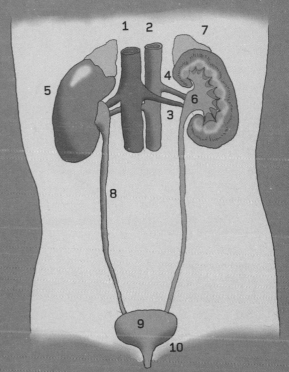

1 Inferior vena cava
2 Aorta
3 Renal vein
4 Renal artery
5 Kidney
6 Nephrons
7 Adrenal gland
8 Ureter
9 Bladder
10 Urethra

Reproductive system

The human reproductive system consists of the organs that allow couples to produce offspring. Male sperm fertilizes a female egg (see page 168), which develops into an embryo, then a full-term baby, during a gestation period of about 40 weeks.

Male testes in the scrotum produce sperm, which mature inside coiled tubules called the epididymis. During ejaculation, sperm travel up through the vas deferens, which loops around the bladder, then out through the penis, with fluids from the prostate and seminal vesicles. Semen contains nutrients for sperm and allows them to 'swim' up to fertilize a woman's egg.

Females are born with their full complement of immature eggs – typically about 2 million – inside their ovaries. Within the ovary, follicles each hold one egg surrounded by cells that nourish and protect it. After puberty, hormones usually make one egg mature each month and travel to the fallopian tubes, where it might be fertilized during sex. A fertilized egg then implants itself in the uterus, which has thick muscular walls and expands as a foetus grows.

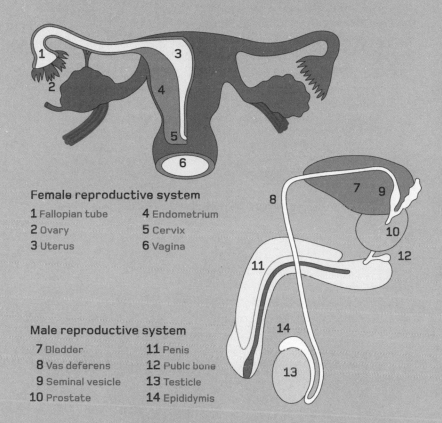

Female reproductive system

1 Fallopian tube
2 Ovary
3 Uterus
4 Endometrium
5 Cervix
6 Vagina

Male reproductive system

 7 Bladder
 8 Vas deferens
 9 Seminal vesicle
10 Prostate
11 Penis
12 Pubic bone
13 Testicle
14 Epididymis

Endocrine system

The human endocrine system is a collection of glands that secrete hormones, which flow through the bloodstream and act as chemical messengers. These molecules trigger chemical changes in cells that have the appropriate receptors (see page 148).

Hormones secreted from the pituitary gland at the base of the brain regulate a host of factors including body growth and temperature, blood pressure, sex organs in both men and women, and some aspects of pregnancy and childbirth. The pineal gland, also in the brain, produces melatonin, a hormone that regulates our sleep patterns.

The thyroid gland controls how quickly the body uses energy and makes proteins. Two adrenal glands release the stress hormone cortisol to trigger a rise in blood glucose levels, while the pancreas secretes insulin, which regulates carbohydrate and fat metabolism. The endocrine system works in tandem with the nervous system (see page 230), which transmits instructions round the body via networks of nerve cells.

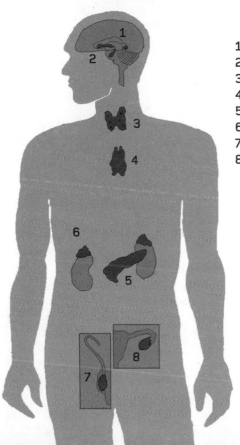

1 Pineal gland
2 Pituitary gland
3 Thyroid gland
4 Thymus
5 Pancreas
6 Adrenal gland (1 of 2)
7 Testis
8 Ovary

Immune system

The immune system is a network of organs, tissues and cells that defends the body against attacks by 'foreign' bodies such as bacteria, viruses, parasites and fungi that can cause disease. It has an amazing ability to track down these pathogens and target them for destruction.

The organs of the immune system include the tonsils, spleen and small bean-shaped lymph nodes laced through tiny lymphatic vessels. They all house lymphocytes, small white blood cells that are the immune system's key players. Immune cells often have specialized functions – they can engulf and digest bacteria, for instance, or kill parasites. They include 'killer T cells', which mature in the thymus and attack tumours and virus-infected cells. Some T cells 'remember' past foes and quickly mount a vicious assault on subsequent encounters.

Unfortunately, immune systems sometimes engage in friendly fire, causing diseases by destroying healthy human tissues. Other problems arise from suppressed immune systems, which can make people vulnerable to diseases such as pneumonia.

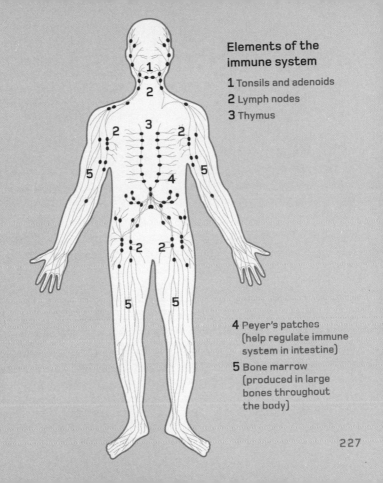

Elements of the immune system

1 Tonsils and adenoids
2 Lymph nodes
3 Thymus
4 Peyer's patches (help regulate immune system in intestine)
5 Bone marrow (produced in large bones throughout the body)

227

Integumentary system

The integumentary system comprises the biggest organ in the human body – the skin – as well as its extensions, such as hair and fingernails. The skin protects delicate organs inside the body and provides a physical barrier to regulate body temperature, keep out foreign bodies and retain moisture.

Most of our skin is about 2–3 mm (0.1 in) thick and it accounts for about 20 per cent of an adult's body weight. The outer layer of the skin is called the epidermis. Its surface consists of dead cells that make skin waterproof, while cell division takes place in the deepest epidermis layer to create new cells that gradually move outwards and replace the outer skin layer.

The dermis lies beneath the epidermis and has its own blood supply, as well as nerves and sweat glands, which collect water and waste products from the bloodstream then excrete them out through pores in the epidermis. Beneath the dermis lies the fatty hypodermis, which connects the skin to underlying bones and muscles.

1 Epidermis
2 Dermis
3 Hypodermis
4 Hair follicle
5 Hair shaft
6 Oil gland
7 Sweat gland
8 Lymph vessel
9 Nerve
10 Fatty tissue

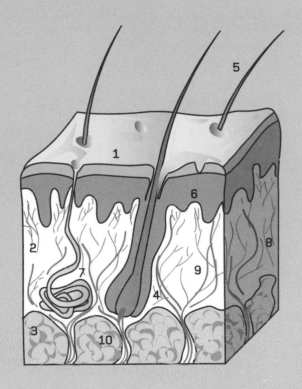

Nervous system

The nervous system is the information highway along which the brain sends instructions and receives feedback. It is made up of billions of nerve cells (neurons) that join together to make nerves, cable-like bundles wrapped in connective tissues, which transmit electrical impulses through the body.

The central nervous system comprises the brain and spinal cord. An adult human brain contains about 100 billion neurons and trillons of 'glia', cells that carry out support functions like transporting nutrients. The spinal cord is a long tubular bundle of nervous tissue that runs down the vertebral column.

The peripheral nervous system extends beyond the central nervous system. It consists of 12 pairs of cranial nerves, which emerge from the brain and mainly serve the head and neck, and 31 pairs of spinal nerves, which branch off from the spinal cord to the rest of the body. The 'autonomic nervous system' is a part of the peripheral nervous system that controls diverse functions from heart rate to the size of our pupils, largely without conscious effort.

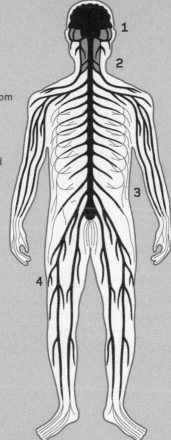

1 Cranial nerves from brain to sense organs, such as eyes, mouth and ears, and other parts of the head

2 Central nerves in brain and spinal cord

3 Autonomic nerves from spinal cord to lungs, heart, digestive system, bladder and sex organs

4 Peripheral nerves connecting spinal cord to limbs

Cardiovascular disease

Cardiovascular disease is a spectrum of disorders that affect the heart and blood vessels (see page 212), including heart attacks and stroke. It is the leading cause of death in developed nations.

Heart attacks occur when a blood clot suddenly blocks an artery in heart muscle. It can cut off most or all of the blood supply to the heart, so heart cells that don't receive enough oxygen-rich blood begin to die. This is often fatal unless a patient receives prompt treatment to restore blood flow.

Strokes occur when the blood supply to part of the brain is cut off and brain cells begin to die, leading to brain damage and possibly death. Most strokes are 'ischaemic' ones, in which a blood clot blocks the blood supply. In 'haemorrhagic' strokes, brain damage occurs when a weakened blood vessel bursts.

The best way to prevent cardiovascular disease is to avoid eating a lot of fatty food, which can make fatty plaques build up in the arteries. High blood pressure and cholesterol, smoking and lack of exercise are also risk factors.

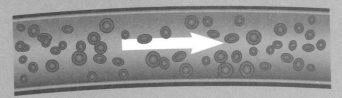

1 Normal blood flow through coronary artery

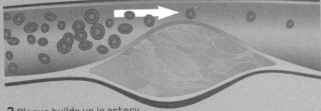

2 Plaque builds up in artery

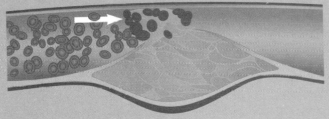

3 Blood flow blocked by artery forms a clot

Infectious disease

Infectious diseases develop when pathogens such as bacteria and viruses (see pages 178 and 182) invade the body. They are a leading cause of death, particularly in developing countries.

Some bacterial infections are beneficial, helping to break down food in our guts during digestion. But harmful bacteria can cause diseases by a variety of mechanisms, including sticking to healthy cells and gumming up their surfaces, and producing toxic chemicals. Fungi can cause diseases such as athlete's foot, while other pathogens include single-celled parasites such as *Plasmodium* that can cause malaria. Many multicellular parasites also cause disease, including tapeworms that can grow several metres long in intestines.

Some rare infectious diseases are caused by 'prions', proteins that have folded into the wrong shape and convert other proteins to the faulty state. Prion diseases destroy the brain, and include bovine spongiform encephalopathy (BSE or 'mad cow disease') in cattle, which can be passed on to people through the food chain as Creutzfeldt–Jakob disease (CJD).

Prion disease cycle

1 Normal protein made in nerve cells
2 Misfolded prion protein
3 Misfolded prion protein infects normal protein in nerve cells
4 New misfolded prions burst out when cell dies

Cancer

Cancers are diseases that develop when cells in the body divide uncontrollably to form lumps called tumours. There are more than 200 types of cancer, and it is the second most common cause of death in developed countries after cardiovascular disease (see page 232).

Tumours can be 'benign' lumps that are harmless, but cancer is a term for 'malignant' tumours with the ability to spread to other parts of the body, either by invading surrounding tissues or by migrating to other organs through the blood or lymphatic system (see page 226). 'Metastasis' occurs when cancer cells reach a new area and continue dividing to create new tumours.

Treatments include surgery to remove malignant tumours and radiotherapy, which destroys them using radiation. In chemotherapy, patients take drugs that target rapidly dividing cells, although this has unpleasant side effects because it also harms healthy cells that normally divide rapidly, including hair follicles. Sometimes, chemicals naturally produced by the body's immune system can shrink tumours with fewer side effects.

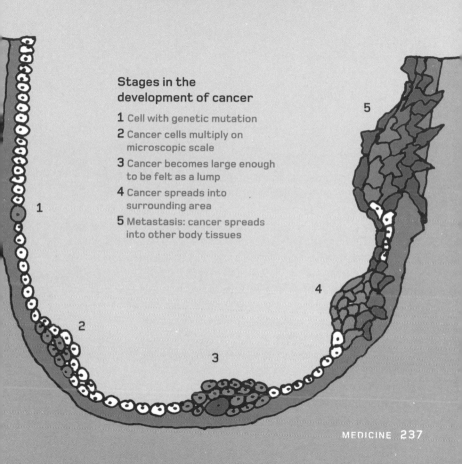

Stages in the development of cancer

1 Cell with genetic mutation
2 Cancer cells multiply on microscopic scale
3 Cancer becomes large enough to be felt as a lump
4 Cancer spreads into surrounding area
5 Metastasis: cancer spreads into other body tissues

Drugs

In general, a drug is any chemical that alters normal body function. Usually, it refers to a chemical designed to treat, cure or prevent disease, or enhance physical or mental health.

Drugs fall into a vast number of classes, including antibiotics that kill bacteria without harming body cells and antiviral drugs that sabotage virus replication strategies. The world's best-selling drug Lipitor lowers cholesterol levels, while other top-selling drugs treat asthma and cardiovascular disease.

'Analgesics' are drugs that relieve pain. Injuries make nerve endings send signals to the brain that trigger pain sensations, and analgesics interfere with these signals in the nervous system anywhere from the injury site to the brain itself. Many painkillers come from naturally occurring chemicals. Aspirin uses a chemical in willow bark, for instance, while opiates work in a similar way to opium, derived from poppies.

Sometimes, people use recreational drugs such as opioids or hallucinogens for their perceived beneficial effects on mood or perception, but many of these are highly addictive.

Stages for drug treatment of pain

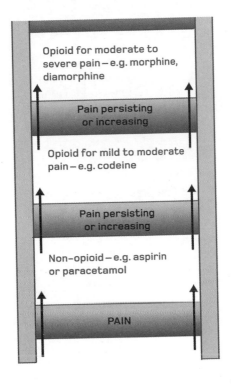

Opioid for moderate to severe pain – e.g. morphine, diamorphine

Pain persisting or increasing

Opioid for mild to moderate pain – e.g. codeine

Pain persisting or increasing

Non-opioid – e.g. aspirin or paracetamol

PAIN

IVF

In vitro fertilization, or IVF, is a technique that allows some infertile women to become pregnant. Doctors might recommend it if a woman has damage to her fallopian tubes (see page 222) or if her partner has a low sperm count.

During the IVF process, the woman usually takes drugs to boost the numbers of mature eggs in her ovaries. Then doctors remove the eggs, usually by guiding a needle into the ovaries monitored by an ultrasound scanner. The eggs are mixed with sperm and cultured in the lab.

If embryos successfully develop, usually one to three of them are implanted into the woman's uterus. More embryos means a higher chance of a successful pregnancy, but many countries have guidelines or laws that restrict the number of embryos used because of the risks of multiple pregnancies, which often lead to premature birth.

Typically, only about a quarter to a third of women become pregnant after one IVF attempt, although the chance of success is highly dependent on the woman's age.

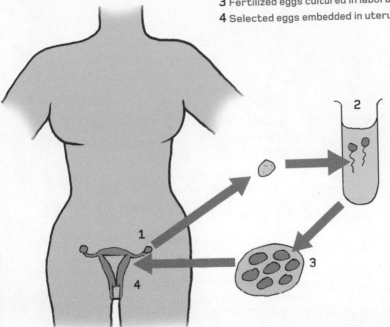

Key steps of in vitro fertilization

1 Eggs extracted from ovaries
2 Eggs fertilized in laboratory
3 Fertilized eggs cultured in laboratory
4 Selected eggs embedded in uterus

Kidney dialysis

Kidney dialysis is a treatment for people who have poor kidney function, usually as a knock-on effect of diabetes or uncontrolled high blood pressure, or due to inflammation. Dialysis carries out key kidney functions by filtering blood to remove waste, salt and excess water.

In 'haemodialysis', blood is drawn out from an artery and pumped into a dialysis unit. Inside the unit, waste products in the blood pass into a fluid called the dialysate through holes in a membrane that are too small to admit blood cells. The cleaned blood is then returned into a vein. Typically, haemodialysis treatments take place three times a week and last about three or four hours.

In 'peritoneal dialysis', blood is cleaned inside the body with the lining of the abdominal cavity acting as the membrane. Dialysate is flowed into the abdomen through a permanent tube and then extracted after the fluid has absorbed waste products and excess water from arteries and veins that line the peritoneal cavity.

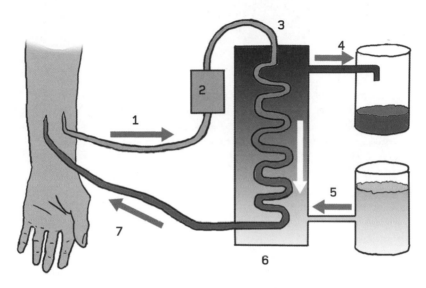

1 Blood drawn from artery
2 Pump
3 Semi-permeable membrane
4 Used dialysate

5 Fresh dialysate
6 Dialysis unit
7 Cleaned blood returned
to vein

Surgery

Surgery is a medical procedure to manually remove or modify tissue in the body, usually to treat disease. Surgical operations began at least 7,000 years ago, when Stone Age people used flint tools to cut open skulls, possibly to treat head injuries or for other perceived health benefits.

Modern surgery takes place in theatres with carefully sterilized surgical instruments. Patients are given local anaesthetics that numb the part of the body surgeons will operate on, or general anaesthetics that make them unconscious. Common surgical operations include caesarean sections to deliver babies through the abdomen and repairs of hernias (which usually involve part of the intestine protruding through a hole or weakness in the wall of the abdomen).

In laparoscopic, or 'keyhole', surgery, surgeons make tiny incisions in the body and perform surgery guided by a miniature camera attached to a long surgical instrument. Keyhole surgery is often used to remove the gall bladder. The smaller incision means less pain, scarring and risk of infection.

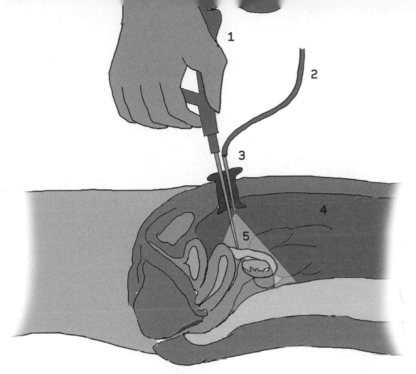

1 Surgical instrument
2 Laparoscope
3 Multi-instrument laparoscopy port

4 Abdominal cavity
5 Illuminated area

Blood transfusion

Blood transfusions involve taking blood from one person – a donor – and giving it to another person. Patients sometimes need transfusions after losing blood due to injury, during operations or in childbirth, or because they have a disease that stops them producing enough red blood cells.

Usually, collected blood is mixed with an anticoagulant as it drains from a catheter in the donor's vein into a plastic bag. Tests determine the donor's blood type, because it has to be compatible with the recipient's blood type, otherwise the patient's immune system will reject it. Blood types have four genetic categories: A, B, AB and O. About 40 per cent of the population are 'universal donors' with O-type blood, which is safe for anyone to receive. Patients with an AB blood type are 'universal recipients', who can safely receive any type of blood.

The blood is also screened for infectious agents including HIV (human immunodeficiency virus) and then ideally separated into its three main components – red blood cells, plasma and platelets – to make best use of it for patients' individual needs.

Blood types

	Group A	Group B	Group AB	Group O
Red blood cell type	A	B	AB	O
Antibodies present	Anti-B	Anti-A	None	Anti-A and Anti-B
Acceptable donors	A or O	B or O	All	O

Laser therapy

In laser surgery, surgeons use a laser to cut or remove tissue instead of a scalpel. They sometimes use lasers to make incisions for otherwise conventional surgery, or to vaporize unhealthy tissues that have high water content. Lasers are sometimes used in cosmetic surgery to destroy the outer skin on the face, to stimulate the growth of new skin that is softer and less wrinkly or scarred.

Laser surgery is commonly used on the eye. Doctors use a laser to vaporize part of the cornea in order to change its shape and correct short-sightedness (myopia) or long-sightedness (hyperopia). Green lasers are often used to shrink enlarged prostate glands in men, with the green light being highly absorbed by the red prostate tissue.

Dentists are increasingly using lasers to replace dental drills for almost painlessly removing decayed parts of teeth, as well as speeding up the bleaching process to whiten teeth. A huge benefit of laser surgery is that there's no physical contact with a surgical instrument, reducing the risk of infection.

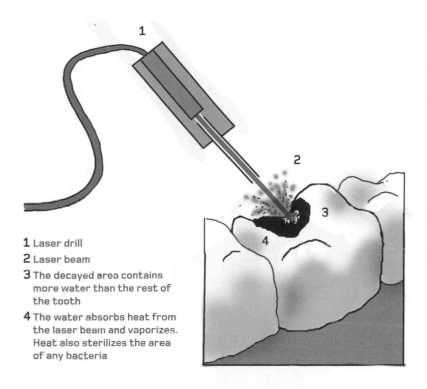

1 Laser drill
2 Laser beam
3 The decayed area contains more water than the rest of the tooth
4 The water absorbs heat from the laser beam and vaporizes. Heat also sterilizes the area of any bacteria

Gene therapy

Gene therapy is a technique for treating diseases caused by defective genes in DNA that produce faulty proteins. So far, this treatment is still in an early experimental stage.

In gene therapy, scientists usually alter a virus genetically to carry a section of normal human DNA. They exploit the fact that some viruses incorporate their own DNA into the human genome as part of their replication strategies. So scientists can dupe a virus into adding a normal gene to human DNA to replace a dysfunctional one. The genetically engineered virus targets cells such as lung or liver cells, then introduces the therapeutic human gene, which starts manufacturing the necessary proteins to restore the cells to a healthy state.

Scientists hope this technique could permanently cure a diverse range of genetic diseases, including haemophilia, a male-only disease in which the blood lacks the normal clotting factors so that even minor injuries can cause dangerous blood loss. However, so far no human gene therapy has conclusively proved to be effective, permanent and safe.

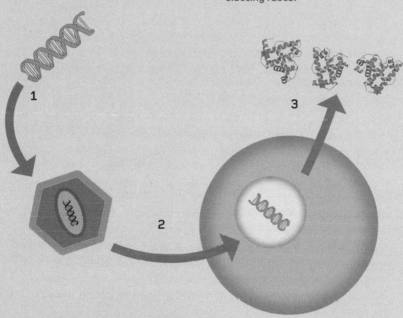

Gene therapy for haemophilia

1 DNA encoding the blood-clotting factor is engineered into a virus

2 The virus introduces the DNA to the human cell nucleus

3 Modified cell produces vital clotting factor

1

2

3

MEDICINE 251

Stem cell therapy

Stem cell treatments could one day cure a wide range of previously incurable diseases, including multiple sclerosis, paralysis and Alzheimer's disease. Found in embryos and various adult tissues including bone marrow, stem cells are unique in their ability to differentiate into a wide range of specific cell types that could be used to regenerate and repair damaged tissue.

Bone marrow transplants are effectively a stem cell treatment for leukaemia. Adult stem cells are limited in the cell types they can generate, but stem cells from embryos can form any type of cell, including liver cells, neurons or skin cells. Treatments using cells derived from human embryos, including neurons for spinal cord repair, are still in the early trial phase.

Scientists hope it will be possible in future to take adult stem cells from a patient needing treatment and programme them to return to an embryonic-like state. These 'pluripotent' stem cells could diversify into any tissue the patient needs, without any risk of tissue rejection by the immune system.

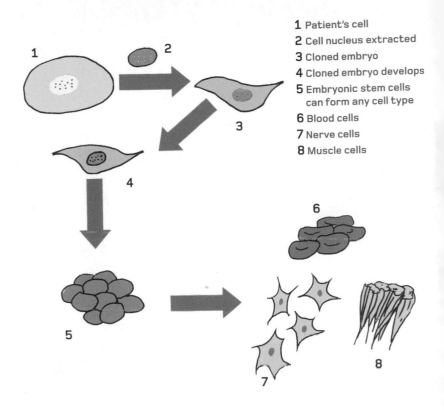

1 Patient's cell
2 Cell nucleus extracted
3 Cloned embryo
4 Cloned embryo develops
5 Embryonic stem cells can form any cell type
6 Blood cells
7 Nerve cells
8 Muscle cells

Earth history

The Earth formed around 4.56 billion years ago, when matter gradually clumped together in a swirling disc of gas and dust around the Sun. The young Earth was hot enough for heavy metals inside to melt and sink into the planet's core, creating a separate core and mantle. About 4.53 billion years ago, a Mars-sized body is thought to have crashed into the Earth, creating the Moon (see page 318).

The Earth's history is divided into four eons, starting with the Hadean, which lasted until 3.8 billion years ago. Towards the end of the Hadean, Earth was pummelled by meteorites during the 'late heavy bombardment'. Water-bearing comets also pelted the Earth's surface, delivering water to form oceans.

Life arose on Earth soon after the late heavy bombardment, and photosynthesis by primitive plants began enriching the atmosphere with oxygen around 3 billion years ago. During the current Phanerozoic eon, covering the last 542 million years, the continents gradually merged into a single landmass called Pangaea, then later split to form the familiar continents today.

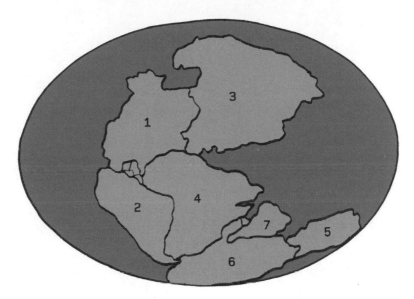

Around 250 million years ago, all Earth's continents were joined together as one giant landmass known as Pangaea (Greek for 'entire Earth')

1 North America
2 South America
3 Eurasia
4 Africa
5 Australia
6 Antarctica
7 India

Earth's structure

The outermost layer of the Earth is the crust, which consists of the continents and the ocean floors. Continental crust is typically about 35–70 km (22–43 miles) thick, while oceanic crust is thinner, typically only about 5–10 km (3–6 miles) thick. The silicate rocks granite and basalt are the most common rocks in the Earth's crust.

The next layer is the mantle, composed mainly of hot, mushy silicates. It is about 2,900 km (1,800 miles) thick. Large convective cells in the mantle circulate heat and drive plate tectonics (see page 264). The Earth has a fluid, iron-rich outer core and a solid inner core, which is probably mostly made of iron with some nickel.

The temperature inside the Earth is thought to rise by about 25–30°C (45–54°F) for each kilometre of depth. Some of that heat is left over from the planet's formation, but most comes from the radioactive decay of unstable elements. Scientists deduce the Earth's deep internal structure by measuring how seismic waves from earthquakes propagate through it.

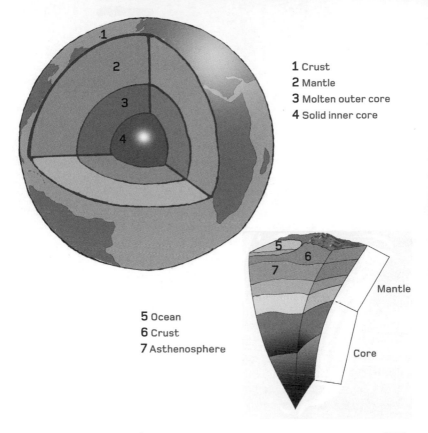

1 Crust
2 Mantle
3 Molten outer core
4 Solid inner core

5 Ocean
6 Crust
7 Asthenosphere

Mantle

Core

Geomagnetism

Geomagnetism refers to the Earth's magnetic field, which is similar to that of a bar magnet. The magnetic north and south poles are close to the geographic poles, but the magnetic poles wander by up to about 40 km (25 miles) each year. The northern and southern lights ('aurorae') are eerie glows that occur near the magnetic poles when energetic particles from the Sun excite atmospheric molecules.

The dynamo theory suggests that Earth's magnetic field sustains itself via a feedback mechanism. The field induces electric currents in the metallic liquid outer core (see page 256), while convection currents and the Earth's rotation organize these currents into spirals aligned from north to south. These currents induce a magnetic field that reinforces the original field, creating a self-sustaining dynamo.

Magnetic fields preserved in ancient lava flows show that the Earth's magnetic field flips over every few hundred thousand years or so, with the north pole moving to the south pole, and vice versa. There is no consensus on why this happens.

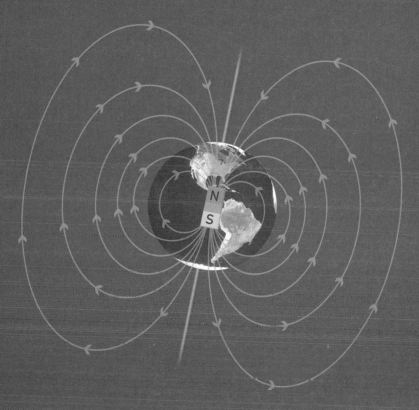

Earth's shape

The Earth's shape is a flattened sphere because it bulges out at the equator slightly due to its rotation. Its average diameter is 12,742 km (7,918 miles), but the polar diameter is about 0.3 per cent less than the equatorial diameter.

The coordinate system for Earth's surface uses lines of latitude and longitude. Longitude lines run north to south, while latitude lines form circles that get smaller towards the poles. By convention, the 'prime meridian' that passes through Greenwich in London marks zero longitude, while zero latitude falls on the equator. The positions of any point on Earth's surface can then be described in degrees north or south and east or west. New York, for instance, is at 41° North, 73° West.

Surveyors and engineers often use the concept of the geoid, a hypothetical Earth surface that represents the mean sea level. It is useful because it represents the horizontal everywhere, and gravity acts perpendicular to it. Water will not flow in an aqueduct, for instance, if its pipes are perfectly aligned along the geoid.

1 Geographic north pole
2 Greenwich Meridian
3 Equator

4 Lines of longitude run
  between poles
5 Lines of latitude run
  parallel to equator

Seasons

Earth's orbit is almost circular, with its distance from the Sun varying by only about 3 per cent over the course of a year. This means that the solar energy received on Earth changes by about 6 per cent. However, this is not the cause of the seasons – hot summers and cold winters are due to the 23.5° tilt of the Earth's rotation axis.

The tilt makes more sunlight fall on the northern hemisphere than the southern hemisphere during the northern summer,

1 Northern summer solstice (June)

2 Equinox: hemispheres receive equal sunlight

3 Southern summer solstice (December)

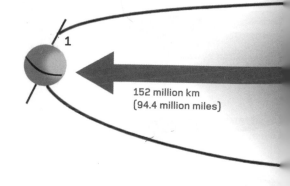

152 million km (94.4 million miles)

the peak occurring on the summer solstice on 20/21 June. More solar energy falls on the southern hemisphere in December, peaking at the solstice on 21/22 December. Sunlight is equal in both hemispheres at the vernal or spring equinox (20/21 March) and the autumnal equinox (22/23 September).

Earth's large axial tilt also means that any regions inside the Arctic and Antarctic circles, at latitudes of more than 66° North or South, will experience a period of permanent sunlight in summer and permanent darkness in winter.

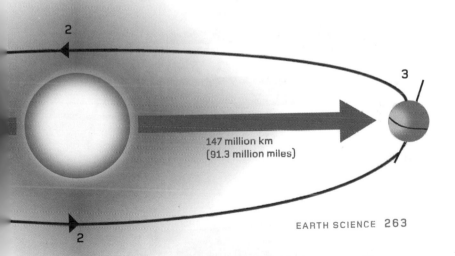

147 million km
(91.3 million miles)

Plate tectonics

Plate tectonics describes the movement of the Earth's lithosphere, consisting of its rigid crust and upper mantle. This is the driving force behind continental drift, which saw a single vast supercontinent called Pangaea break up roughly 250 million years ago, fragmenting to form the familiar modern continents such as Africa and Europe.

The lithosphere divides into several major tectonic plates that move on the mobile mantle underneath. Dense old lithosphere sinks into the deep mantle at 'subduction zones', while new crust is formed by volcanic eruptions at mid-ocean ridges. The speed of tectonic plates is typically very slow – roughly as fast as your fingernails grow.

Where tectonic plates collide, mountain ranges can form, while divergent faults occur when plates move apart. 'Transform boundaries' form where plates are sliding past each other. Earthquakes and volcanoes usually coincide with plate boundaries, although volcanism can also occur at 'hotspots' within plate interiors, which overlie hot mantle plumes.

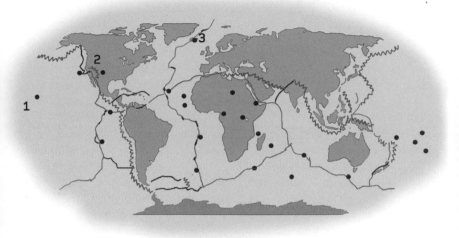

	Divergent plate boundaries
	Transform plate boundaries
~~	Convergent plate boundaries
•	Hotspots

1 Hawaii
2 Yellowstone
3 Iceland

Faults

A fault is a fracture or discontinuity in rocky terrain where two masses of rock have moved relative to each other. Some faults are tiny, but others are part of vast fault systems criss-crossing the Earth at the boundaries of major tectonic plates (see page 264). The sudden movement of faults causes earthquakes. Faults that have horizontal movement are called strike-slip faults, while those with primarily vertical movement are called dip-slip faults.

A divergent fault is one where two plates gradually move apart, sometimes creating mid-ocean ridges as underlying magma wells up through cracks in the oceanic crust and cools. Tectonic plates collide at convergent faults. Sometimes, this makes oceanic crust slide beneath the other plate, forming a 'subduction zone'. The collision of two continental plates can drive up huge mountain ranges like the Himalayas.

A transform fault is one where tectonic plates slide past each other horizontally. A classic example is the San Andreas Fault in California, which has triggered several major quakes.

1 Thick continental crust

2 Volcanic activity
along divergent fault

3 Oceans fill low–lying basins

4 Newly formed oceanic crust

5 Plates pulled apart along
divergent fault

6 Convection currents in
underlying mantle

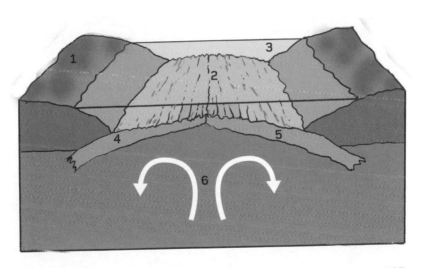

Earthquakes

Earthquakes occur when a sudden release of energy in the Earth's crust shakes the ground by generating seismic waves. They happen because tectonic plates (see page 264) don't glide over each other smoothly without friction. Instead, their roughness makes them lock together, allowing stresses and strains to build up until they lurch sharply.

Divergent faults pulling apart trigger 'normal' earthquakes, convergent plates cause 'thrust' earthquakes and transform faults, where plates slide past each other, cause 'strike-slip' quakes. Traditionally, the power of earthquakes has been measured on the Richter scale, and quakes with 'magnitudes' above nine devastate areas thousands of kilometres across.

When an earthquake occurs under the sea, the seabed sometimes moves enough to trigger tsunamis, giant waves that can devastate coastal regions. An earthquake in December 2004 off the coast of Sumatra, Indonesia, caused the worst tsunamis in recorded history, killing more than 230,000 people in 14 countries.

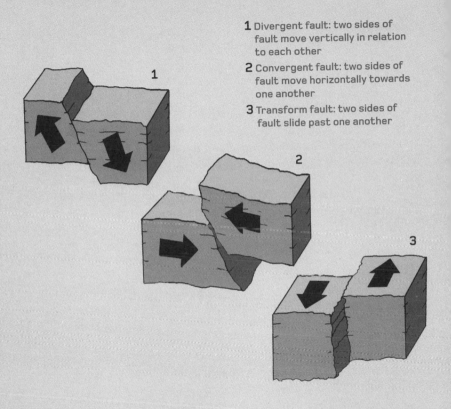

1 Divergent fault: two sides of fault move vertically in relation to each other

2 Convergent fault: two sides of fault move horizontally towards one another

3 Transform fault: two sides of fault slide past one another

Volcanoes

Volcanoes form when hot molten rock, or magma, wells up through the Earth's crust due to heating from the mantle beneath. They're often found along boundaries where tectonic plates (see page 264) converge or diverge – for instance, along the Mid-Atlantic Ridge where plates are pulling apart.

Volcanoes also occur at 'hotspots' far from plate boundaries, where the crust overlies a hot mantle plume. Eruptions at an undersea hotspot formed all of the Hawaiian islands, for instance. Volcanoes often form conical mountains that spew lava, ash and gases from a collapsed crater, or caldera, at the top, but others have rugged peaks formed by lava domes.

'Pyroclastic flows' of searing hot gas, ash and rock often speed away from an erupting vent at up to 150 km/h (90 mph), hugging the ground. Volcanoes also eject volcanic 'bombs', blobs of molten rock up to several metres wide, which cool and crust over before hitting the ground. The most deadly eruption in recorded history was that of Indonesia's Mount Tambora in 1815, which killed at least 71,000 people.

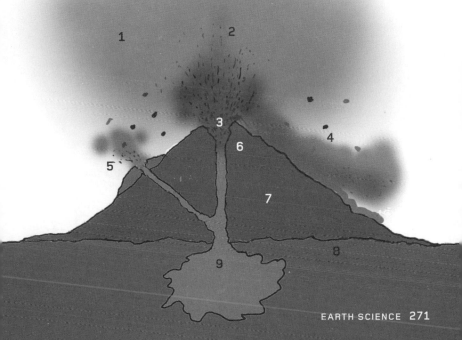

1 Volcanic ash cloud
2 Explosive eruption
3 Caldera
4 Pyroclastic flow
5 Flank eruption
6 Throat
7 Cone of erupted lava
8 Bedrock
9 Magma chamber

Rock types

Geologists classify rocks into three main groups: igneous, sedimentary and metamorphic. Igneous rocks form when hot molten rock, or magma, rises through the Earth's crust, then cools and solidifies. When magma slowly cools deep underground, large crystals grow inside it, creating coarse-grained rock such as granite, while rapid cooling at the surface creates fine-grained rock such as basalt.

Sedimentary rocks form on the Earth's surface. They are layered accumulations of sediments including rock fragments, minerals and animal and plant material. One example is sandstone, which forms when sand settles out of water, then becomes compacted by overlying deposits. Sedimentary rocks probably make up only about 5 per cent of the Earth's crust, forming a thin veneer over igneous and metamorphic rocks.

Metamorphic rocks were once sedimentary or igneous rocks, but their densities increased and their compositions changed when they were pulled deep down into the Earth's crust and subjected to high pressures and temperatures.

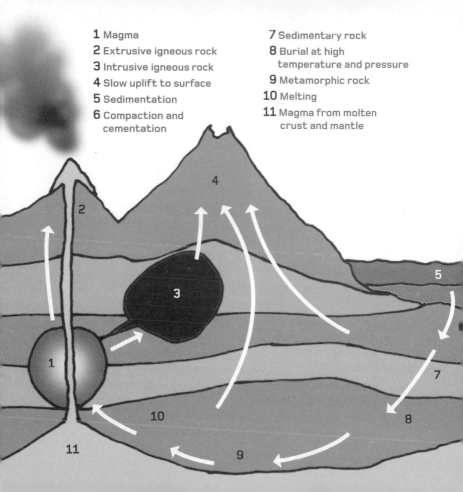

The rock cycle

The rock cycle describes the endless natural recycling processes that rocks undergo on the restless Earth, continually changing over millions of years due to processes such as erosion and tectonic plate motions (see page 264). The rock cycle is particularly active where tectonic plates meet.

The cycle begins with magma, fluid or mushy hot rock beneath the Earth's surface, which cools and crystallizes to form igneous rocks. These rocks can return to their roots as magma by 'subduction', being dragged back down through the crust to melt again. Alternatively, burial of igneous rocks can compress and heat them to form metamorphic rock. At the Earth's surface, rocks are weathered and eroded into fragments and grains. Rivers and streams sweep these particles away and deposit them in lakes and seas, beginning the process of sedimentation that creates sedimentary rock.

Continental crust recycles very slowly, and Earth's current continental crust is typically about 2 billion years old, while the oldest oceanic crust is only about 200 million years old.

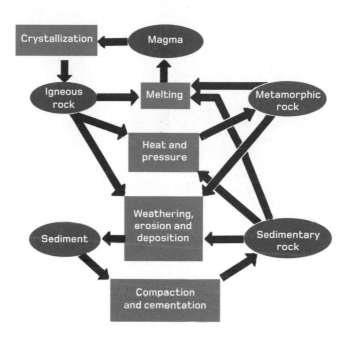

Fossils

Fossils are the remains of animals, plants and other living organisms that have been preserved for thousands of years inside sediments, which have gradually replaced their tissues with minerals.

Fossilization can preserve the remains of animals or plants that are buried soon after they die. For instance, the soft parts of a dead fish might rot away while its skeleton becomes buried in muddy or sandy sediments, retaining its structure as the sediments are compacted into stone. Minerals gradually replace the skeleton by filling voids left as the skeleton slowly dissolves. Millions of years later, this skeleton 'copy' can become exposed through mountain or cliff uplift and erosion.

Like living organisms, fossils can range from microscopic single cells to gigantic dinosaurs and trees. Fossils may also preserve the marks left by animals in sediments, such as footprints of our early human ancestors. The oldest known fossils are 'stromatolites', fossilized colonies of microbes that date back for 3.4 billion years or more.

1 Living organism dies in a sedimentary environment (typically underwater)

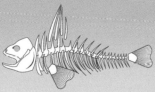

2 Soft tissues decay, but organism is buried before hard parts can be scattered or destroyed

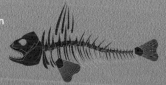

3 Hard parts are buried in sediment, compressed and replaced by minerals over long periods of time

Topography

In geography, topography is the study and mapping of Earth's surface shape and features in three dimensions. Topographic maps, or relief maps, record the height of terrain using contour lines, with each contour line tracking land of equal height. So mountains appear as concentric loops, the steepest slopes indicated by the most tightly packed contours.

Detailed information about terrain and surface features is essential for planning and executing any major projects in civil engineering or land reclamation, for instance. 'Photogrammetry' is a traditional technique for locating 3D coordinates of points on the ground by comparing two or more aerial photos taken from different angles.

Digital data for precise relief maps of the Earth's surface come from satellite radar mapping of the land, while sonar surveys from ships can measure the terrain on the ocean floors. Airborne 'lidar' (Light Detection and Ranging) systems can also map the detailed heights of forest canopies and glaciers, for instance, by measuring reflected visible laser light.

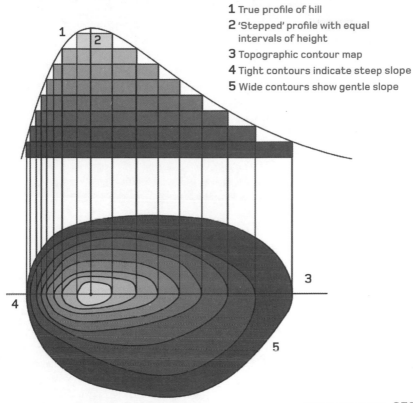

1 True profile of hill
2 'Stepped' profile with equal intervals of height
3 Topographic contour map
4 Tight contours indicate steep slope
5 Wide contours show gentle slope

Continents

The continents are the seven biggest landmasses on Earth: Asia, Africa, North America, South America, Antarctica, Europe and Australia. They make up just over 29 per cent of the Earth's surface. Oceans or seas separate most of the continents, except for Europe and Asia, which are often considered to be a single continent called 'Eurasia'.

Close to 40 per cent of the Earth's total land surface is used for crops and livestock pasture, while roughly a quarter is mountainous. Forests cover about a third of the land. In the tropics, most forests are lush tropical rainforest, with annual rainfall above about 1.8 m (6 ft). Deserts are dry areas with less than 25 cm (10 in) of rainfall each year, making vegetation sparse or almost non-existent. Hot and cold deserts take up about one-fifth of the Earth's land surface.

'Temperate' regions with relatively mild climates lie between the permanently hot tropics and the polar regions, while vegetation-poor 'tundra' with permanently frozen subsoil dominates the ice-free land at high northern latitudes.

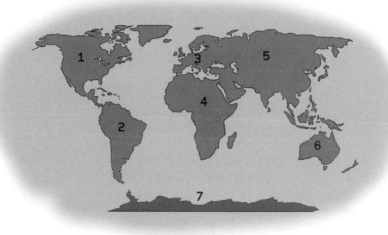

1 North America
2 South America
3 Europe
4 Africa

5 Asia
6 Australia
7 Antarctica

Oceans

The oceans are vast bodies of salt water that cover almost 71 per cent of the Earth's surface. They are usually divided into five major oceans: the Pacific Ocean, the Atlantic Ocean, the Indian Ocean, the Southern Ocean and the Arctic Ocean.

Nearly half of all oceans, by area, are more than 3 km (9,800 ft) deep. The deepest point overall is in the Mariana Trench in the Pacific south of Japan, which reaches down about 11 km (36,000 ft). Two oceanographers, Don Walsh and Jacques Piccard, reached the bottom of the Mariana Trench in a small submersible in 1960 – a feat no one else has achieved since.

Ocean currents act like giant conveyor belts to transfer heat from the tropics to the poles. Cold deep water rises and warms in the central Pacific and the Indian Ocean before heading to high latitudes where it sinks and cools. An important ocean current system stretching from the southeast US to northwest Europe incorporates the Gulf Stream and the North Atlantic Drift, and helps keep northwest Europe's climate relatively warm.

—— Cold currents	**1** Gulf Stream
—— Warm currents	**2** North Atlantic Drift

Surface water

About 97 per cent of all water on Earth is in the salty oceans, while only about 2.5 per cent is fresh water. Most of that is tied up in the ice caps or lies underground. In fact, only about 0.3 per cent of Earth's fresh water is in rivers and lakes, the sources of most water we use in everyday life.

When the Sun heats water in the oceans, it evaporates as water vapour that rises and condenses into clouds, before falling down as precipitation including rain and snow. Ice can be stored for thousands of years in the ice caps and glaciers, which contain about 70 per cent of Earth's fresh water.

Rainwater runs off land into rivers that flow to the oceans or into largely freshwater lakes. There are many types of lakes, including 'oxbow lakes' that form when the force of flowing water gradually exaggerates a meandering curve until it cuts off from the main river channel. Lake Superior on the US–Canada border is often regarded as the largest freshwater lake by area, covering 82,400 square km (31,820 square miles).

All the water in the oceans, rivers, atmosphere and ice caps would form a sphere just 1,390 km (860 miles) across, amounting to roughly 0.13 per cent of our planet's volume – as shown by this sphere of liquid alongside Earth.

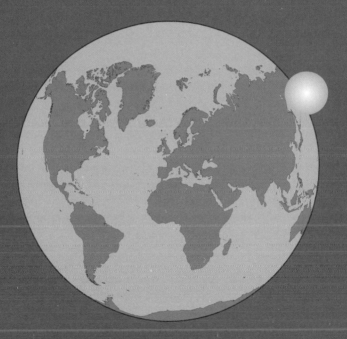

Atmospheric chemistry and structure

The atmosphere is the shroud of gases around the Earth, held in place by gravity. It plays a vital role in making our planet hospitable to life, providing air to breathe and preventing large temperature swings between night and day.

The atmosphere is mainly composed of nitrogen (78 per cent) and oxygen (21 per cent), but its composition changes with height. The lowest layer, the troposphere, is the densest and contains roughly 80 per cent of the atmosphere's mass. The next layer is the stratosphere, and this contains the ozone (O_3) layer, which absorbs most of the ultraviolet light from the Sun that would otherwise be harmful to life. The outermost atmospheric layer is the thin exosphere, composed mainly of hydrogen and helium.

The Earth's atmosphere looks blue because it scatters blue sunlight better than red sunlight, sending blue light photons in every direction. Sunrises and sunsets appear red because the Sun is on the horizon, so its light passes on a long path through the atmosphere and more blue light is removed.

1 Earth's surface
2 Troposphere – up to 10–17 km (6–11 miles)
3 Stratosphere – up to 51 km (32 miles)
4 Mesosphere – up to 85 km (53 miles)
5 Thermosphere – up to 350–800 km (220–500 miles)
6 Exosphere (upper limit undefined)

Atmospheric circulation

The atmospheric circulation is the large-scale movement of air that distributes heat across Earth's surface. It is dominated by 'Hadley cells', huge convection loops described by English lawyer and scientist George Hadley in the early 1700s.

Hadley cell circulation begins with moist, hot air at the equator rising and moving polewards, then descending at latitudes of about 30° North and South. Some of the descending air travels across the surface back towards the equator, creating the 'trade winds' that also veer towards the west due to the Earth's rotation. The polar cells are high-latitude convection loops at more than 60° North and South.

Ferrel cells, first proposed by 19th-century American meteorologist William Ferrel, are convection cells that operate at mid-latitudes but rotate in the opposite direction to polar cells and form westerlies due to the Earth's rotation. The jet streams – mainly the 'polar jets' and 'subtropical jets' – are high-altitude flows of fast-moving air that form at the boundaries between the cells and rotate towards the east.

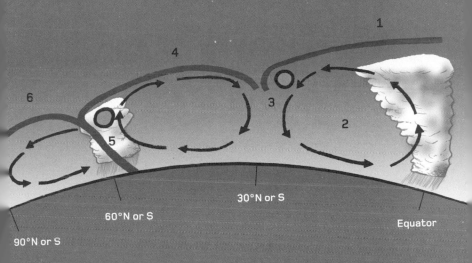

1 Hadley cell
2 Convection cell formed by warm air rising and cool air falling
3 Subtropical jet
4 Ferrel cell
5 Polar jet
6 Polar cell

90°N or S
60°N or S
30°N or S
Equator

Weather fronts

In meteorology, weather fronts are boundaries separating masses of air with different density, temperature and humidity. Their approach signals the onset of a change in weather. For instance, when a cold front moves under a mass of warm moist air, the warm air rises and the moisture can condense into heavy rainclouds.

Cold fronts move faster than warm fronts and produce more sudden changes in weather because cold air is denser than warm air and replaces it rapidly. On weather maps, cold fronts are shown as lines of blue triangles pointing in the direction of travel. Light rainfall often signals the approach of a warm front, depicted as a line with red semicircles.

An 'occluded front' forms when a cold front overtakes a warm front. A 'stationary front' is effectively a stalemate between two fronts, neither strong enough to replace the other. It tends to hang around in the same place for a long time, often delivering rainy weather for several days.

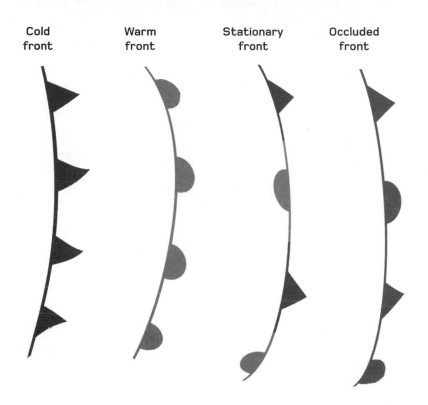

Cold front

Warm front

Stationary front

Occluded front

Clouds

A cloud is an opaque mass of water drops or ice crystals suspended in the atmosphere. Clouds form because sunlight warms the Earth's surface and evaporates water. Moist, warm surface air rises to higher altitudes where the water vapour condenses onto tiny particles like dust or salt, forming liquid droplets or ice crystals if it is cold enough. Eventually, these become too large to be supported by upward air currents and fall as precipitation (see page 294).

Cumulus clouds are puffy, dense clouds that sometimes look like cotton wool. They grow upwards and can develop into giant cumulonimbus clouds that trigger thunderstorms. Cirrus clouds are thin, wispy clouds blown by high winds into long streamers. They form at high altitudes above 6 km (20,000 ft), and usually accompany pleasant weather.

Clouds with the prefix 'alto' are middle-level clouds, while stratus clouds are uniform greyish ones that often cover the entire sky. All weather-related cloud types form in the troposphere, the lowest major layer of Earth's atmosphere.

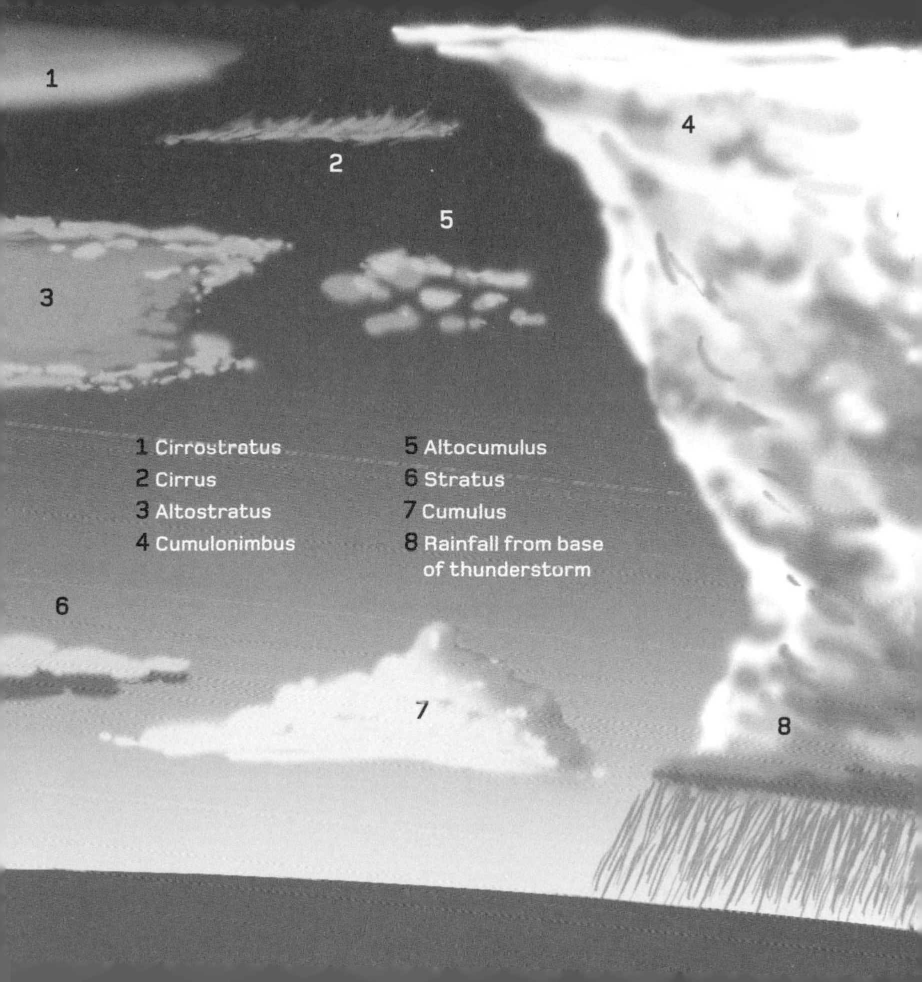

1 Cirrostratus
2 Cirrus
3 Altostratus
4 Cumulonimbus
5 Altocumulus
6 Stratus
7 Cumulus
8 Rainfall from base
  of thunderstorm

Precipitation and fog

Precipitation is any kind of water falling out of clouds, including rain, snow, sleet and hail. It happens when air turbulence inside clouds makes small water droplets or ice particles collide, producing larger ones. When they become too large to be supported by upward air currents, they fall to the ground. (An exception is 'virga', light precipitation that evaporates before it hits the ground.)

Raindrops grow up to about 10 mm (0.4 in) across, the largest ones flattened into pancake shapes by oncoming airflow. Snowflakes can reach several centimetres wide. Hailstones grow as they repeatedly rise and fall inside a cloud by moving in and out of an updraught, and can reach more than 20 cm (8 in) wide, big and heavy enough to cause fatal injuries.

Unlike precipitation, fog is a mass of water droplets or ice crystals suspended in the air at or near the Earth's surface – basically a low-lying cloud. The moisture often has a local source, such as a lake or marsh. Mist is thin fog that allows visibility greater than 1 km (3,280 ft).

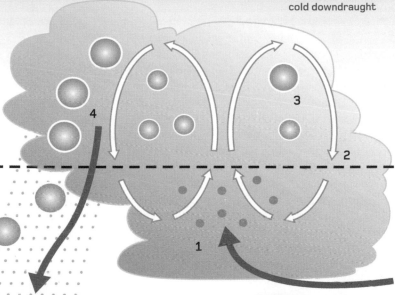

Formation of hail

1 Raindrops sucked into warm updraught

2 Freezing height

3 Hailstones move up and down in convection cells, growing in size

4 Hailstones grow too large to stay in cloud and fall towards ground, creating a strong cold downdraught

Storms and tornadoes

A storm is any disturbance in the atmosphere that causes severe weather. Storms arise when rising hot air creates a centre of low pressure surrounded by high-pressure regions, leading to strong winds and the formation of storm clouds such as cumulonimbus clouds.

Thunderstorms occur in warm regions when humidity is high. Moist, warm air becomes unstable and rapidly rises, while cold air forms strong downdraughts beneath. Falling water drops and ice particles shear negative electric charge off rising ones, causing 'charge separation' in the clouds that discharges in lightning strikes, heard as thunder claps (see page 298).

Tropical cyclones occur at low latitudes when air rotates around a centre of low pressure, fuelled by heat released when moist air rises and condenses. Major tropical cyclones are often called hurricanes or typhoons depending on location. Tornadoes are violent, funnel-shaped wind storms that suck up debris and can persist for more than an hour. They are most common in the central US, in an area dubbed Tornado Alley.

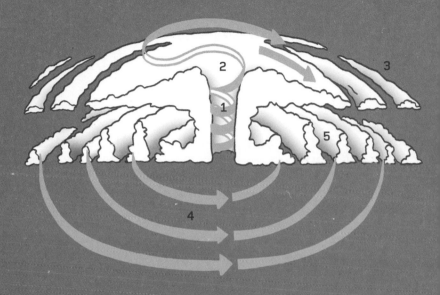

Structure of a tropical cyclone

1 Convection currents
2 Eye
3 Cool, dense air

4 Hurricane winds and rain
5 Warm, moist air

Lightning

Lightning occurs during thunderstorms when electric charge separates inside clouds. As a thunderstorm brews, water droplets in rapidly rising warm air transfer electric charge to falling droplets and ice particles, making the base of a cloud negatively charged relative to the cloud tops. Lightning happens when the resulting electric field becomes powerful enough to discharge through the cloud or to the ground.

Cloud-to-ground lightning starts when a channel of charge, usually negative, zigzags downwards in a forked pattern to the ground and connects with a 'streamer' of positive charge reaching up. This creates a path for a lightning bolt that heats the air, triggering pressure waves that we hear as thunder.

The atmosphere also hosts high-altitude electrical discharges called sprites – usually red luminous glows that sometimes have bluish downward tendrils – and elves – red glows each lasting less than 0.001 seconds. Many people have reported seeing hovering, glowing spheres of 'ball lighting' at ground level, but the origin of this effect is a mystery.

Just before a lightning strike, leaders of negatively charged air (1) start to snake down from the bottom of the cloud, while positively charged leaders (2) snake up from the ground

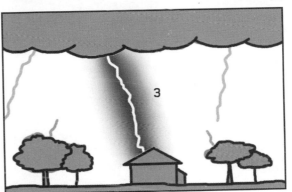

When a leader from the cloud meets a leader from the ground, current can flow along the unbroken pathway (3), and lightning strikes

Climate

The Earth's climate describes regional average weather patterns, including factors such as typical temperature, humidity, wind and rainfall at different times of the year. A host of factors influence these patterns, including latitude, altitude and the location of a landmass relative to an ocean.

The most commonly used climate classification system is the Köppen system, published by German climatologist Wladimir Köppen in 1884. This assigns all regions to five main climate categories. Tropical regions have sea-level temperatures averaging 18°C (64°F) or more all 12 months of the year, while dry climates receive less water from precipitation than they can potentially lose through processes like evaporation.

Temperate regions have seasonal temperatures averaging above 10°C (50°F) in summer and above −3°C (26.6°F) in winter, while continental climates differ by having a coldest monthly average below −3°C. Polar regions have average monthly temperatures below 10°C all year round. These broad categories have 28 subcategories in total.

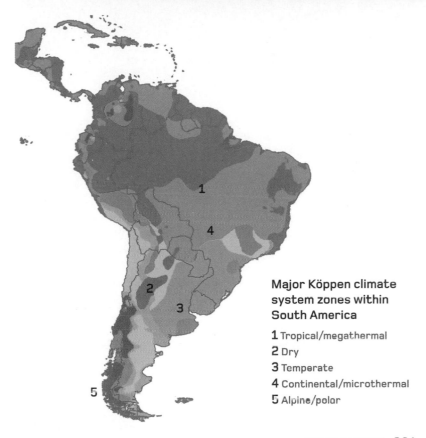

Major Köppen climate system zones within South America

1 Tropical/megathermal
2 Dry
3 Temperate
4 Continental/microthermal
5 Alpine/polar

Climate change

Many factors have altered the Earth's climate over time, including tiny periodic variations in its orbit and the orientation of its spin axis. Throughout much of Earth's history, global average temperatures were more than 5°C (9°F) warmer than today and the poles were ice free. At other times, the world has been plunged into ice ages (see page 304).

Climate change has also occurred in recent times. Between the mid-1500s and the mid-1800s, a period dubbed the 'Little Ice Age', average temperatures were roughly 1°C (1.8°F) cooler than today. One possible reason is that atmospheric ash from volcanic eruptions cooled the planet by blocking sunlight.

Average temperatures increased by 0.6–0.9°C (1–1.6°F) during the 20th century. Most scientists believe this is due to human activity, especially burning fossil fuels. This releases greenhouse gases, which trap some solar energy that would otherwise escape into space. Temperatures could climb by several degrees during the 21st century, causing catastrophic sea-level rise and triggering frequent droughts and storms.

The greenhouse effect

1 Incoming solar radiation
2 Radiation reflected by atmosphere
3 Radiation reflected by Earth's surface
4 Reflected radiation absorbed by greenhouse gases and re-emitted back towards the surface
5 Solar energy absorbed by Earth

Ice ages

In ancient history, the Earth's poles were sometimes ice free, but during ice ages, cool climates allowed vast ice sheets to grow over the continents. There are many natural causes for this continuous climate change, including tiny changes in the tilt of Earth's spin axis and movements of the continents.

Evidence for ice ages comes from geological features such as valleys carved by creeping glaciers as well as deep-drilled polar ice cores, which contain bubbles of ancient air that preserve temperature information. The fossil record shows many organisms spread to warmer regions during cold periods.

There have been at least five major ice ages so far. The earliest well-established one occurred 2.5 to 2.1 billion years ago, while a cold period 850 to 630 million years ago may have seen 'Snowball Earth' conditions with ice reaching the equator. The current 'Quaternary glaciation' began 2.58 million years ago. The Earth is now in an 'interglacial period', a relatively warm period within an ice age, while the last especially cold glacial period ended roughly 10,000 years ago.

During the last glacial of Earth's most recent ice age, glaciers extended across much of northern North America, Europe and Asia, as well as spreading out from the Andes of South America and becoming thicker across Antarctica.

Climate engineering

Climate engineering describes proposed attempts to mitigate global warming on Earth caused by our use of fossil fuels. Each year, fossil fuel consumption releases billions of tonnes of carbon dioxide, a greenhouse gas.

Some climate engineering proposals would reduce the amount of greenhouse gases in the atmosphere directly – for instance, by using industrial plants to mop up the gas, liquefy it and then pump it underground or into the ocean floor. Another idea is to add iron to the oceans, stimulating the growth of ocean phytoplankton that use iron as a nutrient and absorb carbon dioxide as they grow.

Another possible approach is to cool the Earth by cutting down the amount of solar energy reaching the atmosphere. Mirrors on spacecraft could reflect sunlight away, or aircraft could seed the atmosphere with aerosol particles that block out light. These ideas are all in an early research phase – they remain largely unproven as solutions to global warming and could ultimately do more harm than good.

Climate engineering techniques

1 Cloud seeding
2 Giant reflectors in orbit
3 Aerosols in stratosphere
4 Tree planting
5 Greening deserts

6 Iron fertilization of sea
7 Liquid carbon dioxide pumped into rocks
8 Liquid carbon dioxide pumped into deep ocean

Fossil fuels

Fossil fuels, including oil, coal and natural gas, are energy-rich fuels formed underground by the decomposition of dead organisms and plants. They are non-renewable resources because people are extracting and burning them much faster than new ones are forming. Dwindling oil reserves might make oil drilling economically unviable beyond about 2050.

Fossil fuels formed gradually over millions of years as drifting organisms like animals, plants and algae in seas and lakes settled to the bottom and decomposed. This organic matter, mixed with mud, sank under ever deeper layers of sediment until pressure and heat chemically transformed them into liquid and gaseous hydrocarbon molecules. On land, the decomposition of plants tended to form coal and methane.

Fossil fuels are the world's primary source of energy. Each year, fossil fuel burning releases billions of tonnes of carbon dioxide into the atmosphere. This greenhouse gas contributes to global warming, which could unleash devastating climate change in future (see page 302).

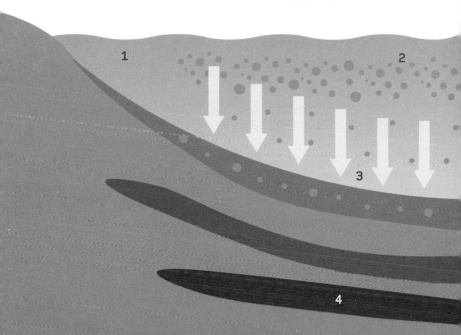

1 Ocean or lake
2 Plankton-rich water

3 Plankton dies and
 falls into sediment
4 Buried muds slowly
 turn into oil

Oil refining

Oil refining is the process of separating the hundreds of different hydrocarbon molecules in crude oil into useful chemicals, including vehicle fuels and lubricating oil as well as ingredients for plastics and detergents.

Crude oil contains hydrocarbons with various molecular masses that are separated into 'fractions' with different boiling points by distillation. The crude oil is heated and its vapour rises through a tall cooling tower, which is coolest at the top. Different fractions condense at different heights in the tower, depending on their mass, with relatively light distillates like petrol fuel condensing near the top and sticky bitumen for roads and roofs gathering at the bottom.

Oil refineries also 'crack' some of the long-chain heavier distillates into lighter, shorter-chain hydrocarbons that are more in demand. For instance, heat and catalysts can break butane into hydrogen and alkenes, important chemicals for the manufacture of polymers (see page 130). Oil refineries can process up to several thousand barrels of crude oil a day.

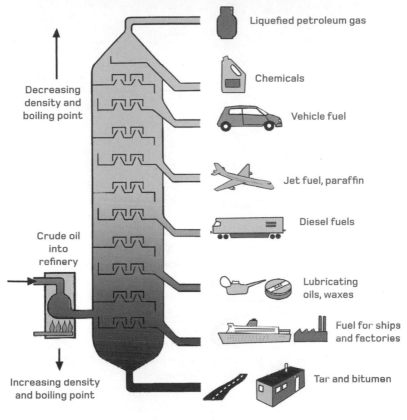

Decreasing density and boiling point

Crude oil into refinery

Increasing density and boiling point

Liquefied petroleum gas

Chemicals

Vehicle fuel

Jet fuel, paraffin

Diesel fuels

Lubricating oils, waxes

Fuel for ships and factories

Tar and bitumen

Nuclear power

Nuclear power is energy generated by controlled nuclear fission reactions. Most nuclear reactors use uranium-235 as a fuel. Neutrons split the uranium atoms, releasing more neutrons and splitting more uranium in a chain reaction that produces heat. Flowing water carries the heat away to generate steam, which turns turbines to generate electricity.

About 14 per cent of the world's electricity comes from nuclear power, and small reactors power some submarines and icebreaking ships. Several serious accidents have occurred at nuclear reactors, including the Chernobyl disaster in the Ukraine in 1986, when a reactor ruptured and caught fire, spewing radioactive fallout over a vast area. These accidents are unlikely in modern reactor designs, but storing dangerous radioactive waste from fission reactors is an ongoing problem.

Reactors using nuclear fusion (see page 98) would produce much less hazardous waste, but are still in the experimental phase. Commercial fusion reactors would have to operate at temperatures of about 100 million °C (180 million °F).

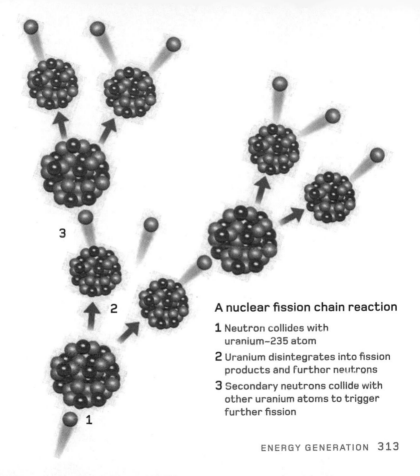

A nuclear fission chain reaction

1 Neutron collides with uranium–235 atom

2 Uranium disintegrates into fission products and further neutrons

3 Secondary neutrons collide with other uranium atoms to trigger further fission

Renewable energy

Renewable energy is generated from natural resources that are endlessly replenished – as opposed to fossil fuels, which take millions of years to form. Concerns about global warming from fossil fuels and high oil prices are driving up demand for renewable energy, which currently generates about one-fifth of the electricity used worldwide.

Renewable energy includes electricity from solar panels, often for individual buildings, as well as electricity from power stations that use wind or natural water flow to turn turbines. Biofuels are fuels produced from organic matter including plants like corn or wheat as well as vegetable oils and animal fats, wood and straw. The US plans an annual production of 36 billion gallons of biofuel – mainly ethanol and biodiesel – by 2022.

Iceland generates all of its electricity from renewable sources, including geothermal power. Hot magma is close to the surface under Iceland. Power stations pump cold water down into the ground where it heats and returns to the surface as steam, which drives turbines to generate electricity.

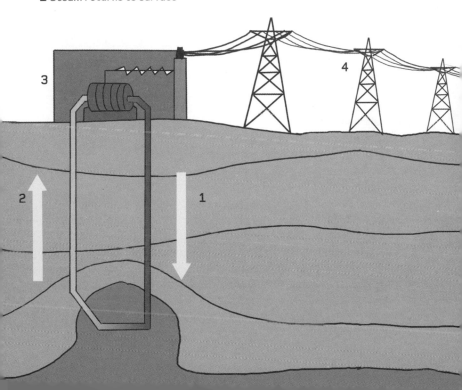

A geothermal power plant

1 Cold water is pumped downwards and heated in hot rocks below surface

2 Steam returns to surface

3 Geothermal steam drives turbine

4 Electricity supplied to grid

The Sun

The Sun is the star at the centre of our solar system. It lies about 150 million km (93 million miles or 8.3 light minutes) away from Earth and has a diameter of 1,391,000 km (864,300 miles). The Sun's composition is almost three-quarters hydrogen, roughly one-quarter helium (by mass), while heavier elements make up less than 2 per cent.

The Sun generates energy by nuclear fusion of hydrogen in its core. Heat moves out to the 'photosphere', where the sunlight we see originates. Beyond that a thin 'corona' expands outwards to form the solar wind, a stream of particles that constantly blows out into space. Sunspots are temporary, relatively cool patches on the Sun where magnetic fields have suppressed heat transfer to the surface.

The Sun formed from a collapsing gas cloud about 4.57 billion years ago. Around 5 billion years from now, it will expand into a red giant star, its outer layers engulfing the planets Mercury and Venus, and possibly the Earth. Eventually, it will shrink into a hot and dense white dwarf.

Structure of the Sun

1 Core generates energy through nuclear fusion

2 Radiative zone transports energy by radiation

3 Convective zone transports energy through convection

4 Photosphere where gas becomes transparent

5 Superhot outer atmosphere or corona

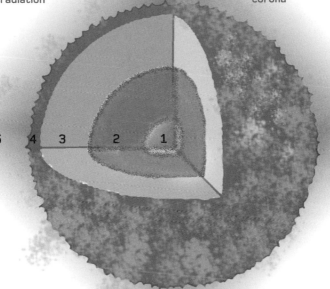

The Moon

The Moon is Earth's only natural satellite. It lies 384,400 km (238,900 miles) away from the Earth on average, although the distance varies by about 5 per cent during the Moon's 27.3-day orbit. The Moon has one-eightieth of the mass of the Earth. Viewed from Earth, it goes through phases as reflected sunlight makes different portions of it visible.

Our satellite is thought to have formed about 4.53 billion years ago, when a Mars-sized body smashed into the newborn Earth, spewing hot debris out into Earth orbit. The debris subsequently clumped together into the Moon, which gradually cooled. Today, it has a layered interior structure, probably with a small, partially fluid core.

Over time, the Earth's gravitational pull on the Moon has forced it into 'synchronous rotation', rotating once for every 27.3-day orbit so that one side faces permanently towards Earth. The surface is pockmarked with millions of craters, more than 5,000 of which are larger than 20 km (12 miles) across. Most formed from the impact of comets and asteroids.

Phases of the Moon

1 New Moon
2 Crescent
3 First quarter

4 Waxing gibbous
5 Full Moon
6 Waning gibbous

7 Last quarter
8 Decrescent

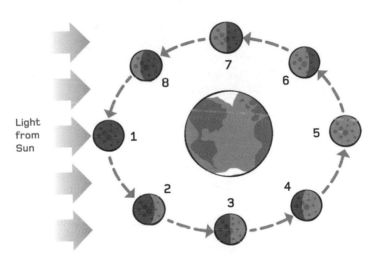

Light
from
Sun

Eclipses

Eclipses are astronomical events that occur when one body passes in front of another and blocks out its light. The most spectacular ones are total solar eclipses, which occur when the Moon lines up with the Sun, as viewed from Earth. The Moon blocks out the sunlight, briefly turning day into night.

Chance alignments between Earth, Moon and Sun create up to two total solar eclipses each year, visible from limited regions of Earth's surface. Because the Sun and Moon appear the same size in Earth's skies, the Moon can obscure the Sun for several minutes. Partial solar eclipses occur when the Moon blocks only part of the Sun.

Total lunar eclipses occur when the full Moon moves into the Earth's shadow and is no longer illuminated by direct sunlight. The Moon appears dark red due to some sunlight reaching the lunar surface after refracting, or bending, through the Earth's atmosphere. The word 'eclipse' can also refer to the apparent coincidence of more distant bodies – for instance, one star briefly blocking the light of a companion star orbiting around it.

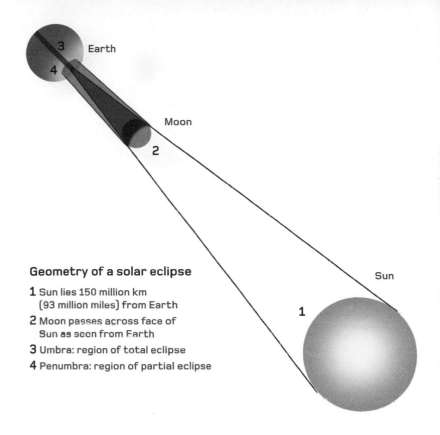

Geometry of a solar eclipse

1 Sun lies 150 million km
(93 million miles) from Earth

2 Moon passes across face of
Sun as seen from Earth

3 Umbra: region of total eclipse

4 Penumbra: region of partial eclipse

Earth

Moon

Sun

Planets

Pluto

The solar system has eight planets: the four terrestrial planets Mercury, Venus, Earth and Mars, and the giant planets Jupiter, Saturn, Uranus and Neptune. They all formed about 4.54 billion years ago, when material clumped together in a disc of gas and dust around the Sun.

Rocky terrestrial planets formed in the warm inner solar system, which favoured compounds with high melting points such as metals and silicates. The giant planets lie beyond the 'frost line', where volatile compounds formed ices that clumped into larger balls capable of capturing heavy atmospheres.

The orbital distances of the planets are measured in astronomical units (AU), where 1 AU is the Earth–Sun distance. A simple numerical relationship called Titius-Bode law predicts the orbit distances. It starts with 0 followed by the doubling-number sequence 3, 6, 12 etc, then

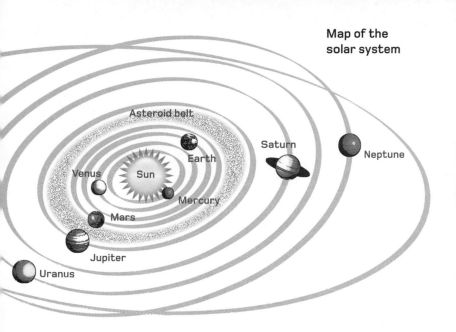

**Map of the
solar system**

adds four to each and divides by ten. The resulting sequence
closely matches the planetary orbit distances (with the
exception of Neptune), but there's no physical reason for this –
it's just a coincidence.

Terrestrial planets
Mercury to Mars

Mercury is the planet closest to the Sun. It orbits the Sun every 88 days and rotates very slowly so that a Mercury day – the time between one sunrise and the next – is 176 Earth days. Temperatures during the long days can climb to 450°C (849°F) on the planet, which has almost no atmosphere, while night chills the surface to –170°C (–274°F).

Venus is the second planet, orbiting the Sun in about 225 days. It's similar in size to Earth but is often described as the Earth's 'evil twin'. It has a crushing, heavy atmosphere of carbon dioxide, a greenhouse gas, which bakes the surface to 465°C (869°F), as well as dense clouds of sulphuric acid.

The Earth is the Sun's third planet, followed by Mars, which takes 687 days to orbit the Sun. Today, the average temperature on this chilly planet is about –60°C (–76°F) and the atmosphere is thin and dry, but there are large deposits of ice buried below the surface, and ancient surface features such as river beds suggest that Mars was once warm enough to have water, oceans and flowing rivers.

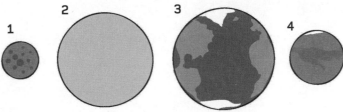

The inner planets

1 Mercury
Diameter: 4,878 km
(3,030 miles)
Year: 88 Earth days

2 Venus
Diameter: 12,100 km
(7,516 miles)
Year: 225 Earth days

3 Earth
Diameter: 12,742 km
(7,915 miles)
Year: 365.25 days

4 Mars
Diameter: 6,794 km
(4,220 miles)
Year: 1.88 Earth years

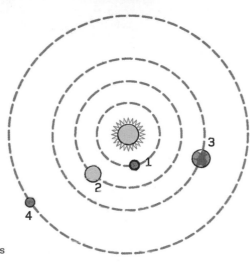

Outer planets
Jupiter to Neptune

The four outer planets of the solar system are vast worlds that together contain almost 99 per cent of all the matter orbiting the Sun. The largest is Jupiter, which is more than 11 times wider than the Earth. Jupiter orbits the Sun every 11.9 years and is famous for colourful banded clouds and the Great Red Spot, a giant storm that has persisted for at least two centuries. Jupiter has dozens of moons including Ganymede, the largest moon in the solar system.

Saturn, like Jupiter, is a gas giant mostly composed of hydrogen and helium. It orbits the Sun every 29.5 years and sports the most magnificent example of an orbiting ring system, packed with ice chunks, some as large as a bus.

Beyond Saturn lie Uranus and Neptune, with orbital periods of 84.3 and 164.8 years respectively. They are often classed as ice giants because they are richer in ices such as water and ammonia than the gas giants. Uranus's rotation axis has a curiously high tilt, so it effectively rotates 'on its side' compared to Earth.

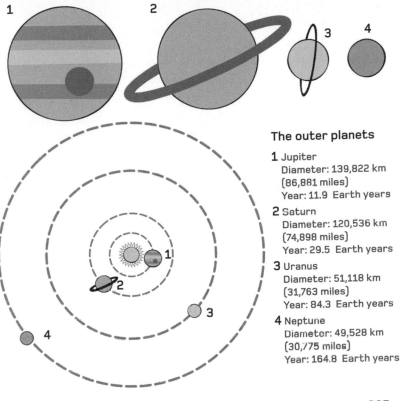

The outer planets

1 Jupiter
Diameter: 139,822 km
(86,881 miles)
Year: 11.9 Earth years

2 Saturn
Diameter: 120,536 km
(74,898 miles)
Year: 29.5 Earth years

3 Uranus
Diameter: 51,118 km
(31,763 miles)
Year: 84.3 Earth years

4 Neptune
Diameter: 49,528 km
(30,775 miles)
Year: 164.8 Earth years

Dwarf planets, asteroids and comets

Broadly speaking, a dwarf planet is a medium-sized world roughly 2,000 km (1,200 miles) wide that orbits a star. The term's precise definition is convoluted, but in our solar system, five known bodies qualify as dwarf planets. They include Pluto, which was classed as a planet until it became clear that there are many similar-sized bodies in the outer solar system; the category 'dwarf planet' was introduced in 2006 to unite them.

Asteroids are smaller than dwarf planets. These rocky chunks mainly circle in the 'asteroid belt' between Mars and Jupiter, although a few have elongated orbits, some crossing the orbit of Earth. Astronomers carefully monitor them to find out if they risk striking the Earth in future, perhaps even causing a mass extinction (see page 204).

Comets are big, dusty snowballs that venture in towards the Sun from two chilly reservoirs of icy bodies – the 'Kuiper belt' beyond Neptune and the more distant 'Oort cloud'. As they approach the Sun and heat up, comets sprout fuzzy atmospheres of gas and dust, and sometimes a long tail.

The outer solar system

1 Sun
2 Orbits of planets
3 Kuiper belt
4 Inner Oort cloud
5 Outer Oort cloud

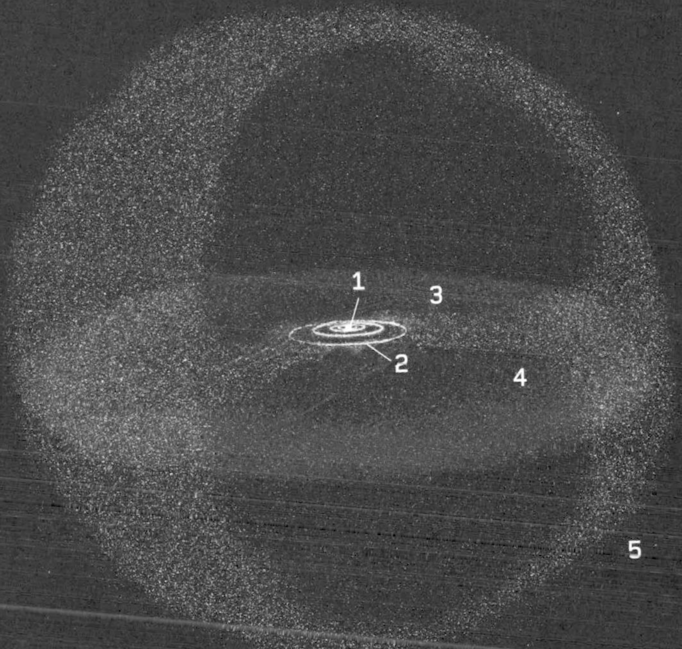

ASTRONOMY 329

Heliosphere

The heliosphere is a vast bubble carved out in space by the solar wind. This bubble envelops all the solar system planets, with its outer boundary marking the region where the solar wind 'loses its puff' and interstellar space begins.

The solar wind (see page 316) blows past all the planets at supersonic speeds of more than 1 million km/h (621,000 mph) before slowing down as it encounters resistance from interstellar gas. The point where it slows below its speed of sound is called the termination shock. Two NASA spacecraft, Voyager 1 and 2, crossed this shock at distances of about 94 and 76 astronomical units (1 AU is the Earth–Sun distance). The shock is probably irregularly shaped and constantly moving.

Beyond this lies the 'heliopause', the theoretical boundary where the interstellar medium brings the solar wind to a halt. Voyager 1 is expected to cross the heliopause by 2014. And beyond this is the 'bow shock', where the interstellar medium hits the outer heliosphere at high speed due to the Sun's orbital motion around the Milky Way.

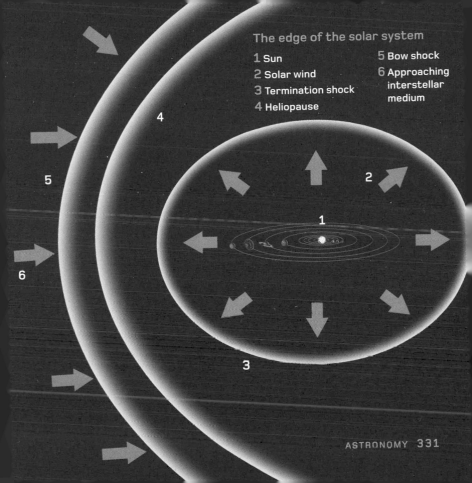

The edge of the solar system

1 Sun
2 Solar wind
3 Termination shock
4 Heliopause
5 Bow shock
6 Approaching interstellar medium

Measuring star distances

German scientist Friedrich Bessel made the first accurate measurement of a star's distance from Earth in 1838, using a technique called parallax. The apparent position of a nearby star in the night sky will vary slightly at two times, six months apart, because the Earth has shifted by about 300 million km (186 million miles) as it orbits the Sun.

Bessel measured the angular shift of a star called 61 Cygni over six months and calculated its distance (about 9.8 light years) by triangulation. Modern satellite measurements have allowed astronomers to calculate the distance of more than 100,000 stars using the parallax method.

More distant stars need another tack. Some variable stars act as 'standard candles' – the timing of their brightness variations reliably indicates their intrinsic brightness, so the *apparent* brightness reveals the distance. Another technique examines the colour of light from distant galaxies. The more distant the galaxy, the more its light will be stretched to long wavelengths by the universe's expansion since the Big Bang.

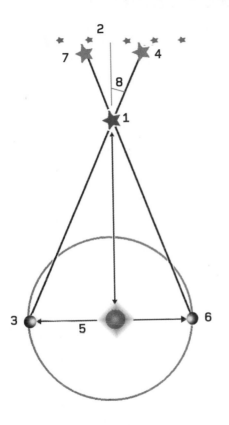

Using parallax to measure the distance to stars

1 Foreground star close to the solar system

2 Distant background stars

3 Earth at time of first measurement

4 Apparent position of star in first measurement

5 Opposite sides of orbit 300 million km (186 million miles) apart

6 Earth at second measurement, six months later

7 Apparent position of star in second measurement

8 'Parallax angle' reveals distance to foreground star

Stellar evolution

Stellar evolution describes the way stars change as they age. Stars form when clouds of gas collapse under their own gravitational pull, and the biggest factor in a star's destiny is its mass. The most massive stars live fast and die young, blowing up in supernova explosions (see page 336) after only a few million years, while theoretically the smallest ones can shine for hundreds of billions of years.

The Sun is a medium-mass star, which will live for about 10 billion years. It is about halfway through its lifespan, spent mostly in a phase called the 'main sequence', during which it generates energy through hydrogen fusion in its core.

Astronomers chart the main stages of stellar evolution on the Hertzsprung–Russell diagram, which plots the colour of a star against its magnitude or luminosity (a measure of its brightness) to reveal patterns. Stars can end their lives in different ways. The Sun will end up as an extremely dense white dwarf, a hot ball of matter roughly the size of the Earth that will gradually cool and fade.

Stellar groups on the Hertzsprung–Russell diagram

1 Main sequence of stars burning hydrogen in their cores – position along line depends on mass of star

2 Red and orange giants are expanding bright stars near the end of their lives

3 Most massive stars swell into supergiants as they age

4 White dwarfs are hot but faint cores left behind when Sun-like stars burn out

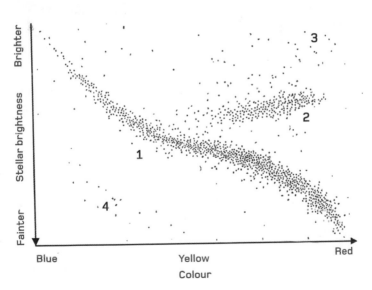

Supernovae

A supernova is a brilliant explosion in which a star blows itself to smithereens. 'Core-collapse supernovae' signal the death of stars more than eight times as massive as the Sun. Fusion reactions gradually build up heavy elements in their cores, but when the fuel runs out, there isn't enough outward pressure to prevent the core suddenly collapsing, sometimes into a black hole (see page 346). This triggers an outward shock wave that catastrophically blows the star's atmosphere apart.

A related phenomenon is the gamma-ray burst, a powerful blast of gamma rays that satellites have detected since the 1960s. Most of these bursts are thought to signal extremely massive, rapidly rotating stars collapsing into black holes.

'Type Ia' supernovae are another main supernova class. They occur when a small, dense white dwarf star grows more massive, either because a companion star 'feeds' it with matter, or because two white dwarfs merge. When the total mass reaches about 1.38 times the mass of the Sun, the star becomes unstable and collapses with a huge release of energy.

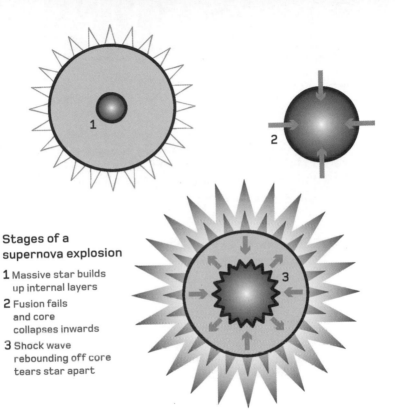

Stages of a supernova explosion

1 Massive star builds up internal layers

2 Fusion fails and core collapses inwards

3 Shock wave rebounding off core tears star apart

Extrasolar planets

An extrasolar planet, or exoplanet, is a world that circles a star beyond our Sun. More than 500 exoplanets have been discovered in our galaxy since the mid-1990s, suggesting that other planets are common throughout the universe.

Most of these worlds have been detected by the 'radial velocity' technique. Astronomers use the Doppler effect (see page 42) to test whether a star is wobbling back and forth due to the gravity of an invisible orbiting planet. Other planet-hunting techniques include the 'transit' method – looking for a slight dimming of a star when a dark planet passes in front of it. A handful of exoplanets have been imaged directly.

Many exoplanets found so far are very unlike the planets of the solar system. Some are 'hot Jupiters', giant planets that zoom round their stars in just a few days, others are 'super-Earths', rocky worlds several times as massive as Earth. Surprisingly, some exoplanets orbit neutron stars (see page 348). The holy grail is to find potentially habitable planets similar to Earth orbiting 'normal' stars like the Sun.

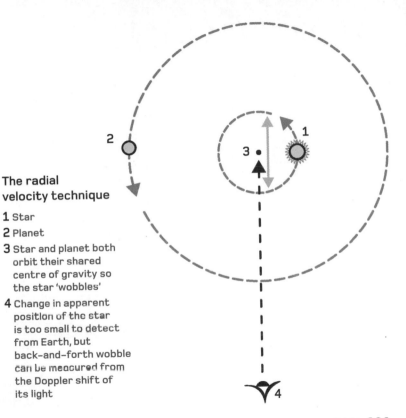

The radial velocity technique

1 Star

2 Planet

3 Star and planet both orbit their shared centre of gravity so the star 'wobbles'

4 Change in apparent position of the star is too small to detect from Earth, but back-and-forth wobble can be measured from the Doppler shift of its light

The Milky Way

The Milky Way is the galaxy of stars that hosts our own solar system. It's a magnificent example of a large spiral galaxy, containing roughly 400 billion stars.

Most of the Milky Way's stars lie within a structure shaped like two fried eggs back to back. A vast disc of stars (the egg whites), roughly 100,000 light years wide, has a central starry bulge (the yolks) with a supermassive black hole (see page 346) at its centre. The disc has several bright spiral arms where dense gas fuels vigorous star formation. Our own solar system lies in the disc, about 26,000 light years away from the galactic centre, which it orbits every 230 million years.

The Milky Way's disc is surrounded by a large spherical halo containing old stars and tight-knit balls of stars called globular clusters, while the whole galaxy is embedded in a vast cloud of invisible dark matter (see page 360). Sometimes, the term 'Milky Way' is used to mean the dense band of stars that crosses the sky where we look across the plane of the galactic disc.

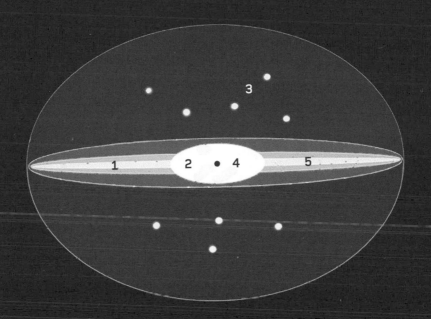

Features of the Milky Way galaxy

1 Galactic disc
2 Galactic bulge
3 Halo containing globular
   star clusters

4 Supermassive black hole
5 Position of solar system

Galaxy types

Galaxies are groups of millions or billions of stars bound together by their mutual gravitational pull. They also contain interstellar gas and dust, as well as vast quantities of dark matter (see page 360).

Galaxies come in three main types. Spiral galaxies, including our home galaxy the Milky Way, have a disc of stars threaded by spiral arms where vigorous star formation takes place.

The Hubble 'tuning fork' model of galaxy classification

1 Elliptical galaxies are classified from E0 to E7 depending on how spherical or flattened they are

2 Lenticular galaxies have a spiral–like hub and disc, but no spiral arms

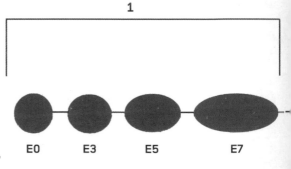

E0   E3   E5   E7

Elliptical galaxies, which include the most massive known galaxies, are spherical or oval in shape. Galaxies that don't have a spiral or elliptical shape are classified as 'irregular'.

Galaxies frequently collide, their gas and dust sometimes combining to trigger vigorous star formation in a new 'starburst galaxy'. Active galaxies (see page 344) emit enormous amounts of radiation, but most galaxies are dim dwarfs containing less than a few billion stars. Galaxies mill around each other in clusters bound by their mutual gravitational attraction, while clusters congregate in 'superclusters' spanning several hundred million light-years.

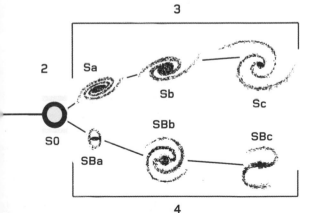

3 Normal spiral galaxies are classified from Sa to Sc according to the tightness of their spiral arms

4 Barred spiral galaxies, classified from SBa to SBc have a bright straight bar of dense stars in the centre

Active galaxies

Active galaxies have bright cores that emit amazingly large amounts of radiation, outshining all their stars. They are so bright that we see them at enormous distances, sometimes so far away that their light has taken more than 13 billion years to travel to Earth.

They are thought to contain supermassive black holes in their cores, which feed on stars and interstellar gas spiralling towards them in a swirling disc. This matter becomes searingly hot as it swirls inwards, so that it emits extremely intense radiation. At the same time, two energetic jets of particles emerge perpendicular to the disc, blasting out across thousands of light-years of space.

Active galaxies fall into different categories, including quasars, Seyfert galaxies and blazars, which have different patterns of light emission. However, astronomers suspect that they're all similar objects viewed from different angles. For instance, blazars are probably the subset of active galaxies that have one of their jets pointed directly towards the Earth.

1 Surrounding 'host' galaxy

2 Central black hole draws in material

3 Acretion disc of superhot material
around black hole forms quasar

4 Jets of material escape above
and below quasar

5 High-energy particles in jets
emit bright radiation

Black holes

Black holes are dark voids in space from which nothing – not even light – can escape. A black hole can form when a very massive star dies in a supernova explosion (see page 336), leaving behind a dense core so heavy that it can't support its own weight. It collapses to a tiny point of enormous density with an immense gravitational pull.

Black holes have a theoretical boundary around them called the event horizon, which marks the point of no return. The size of a static black hole's event horizon is proportional to its mass. The dark, inescapable region of a 10-solar-mass black hole would be roughly 60 km (37 miles) wide.

Much heavier black holes lurk at the centres of large galaxies. They have masses thousands to billions of times higher than the Sun, but it's unclear how they formed. Possibly, many smaller black holes merged. No one can see black holes because they don't emit light, but astronomers can sense their presence by watching their gravitational influence on nearby stars and detecting the radiation from infalling gas and dust.

Light passing close to a black hole experiences an inward pull similar to that experienced by a boat near a whirlpool:

1 Light is moving fast enough to overcome gravity and escape

2 Event horizon

3 Inside event horizon gravity is so strong that light cannot escape

4 On the edge of the event horizon, light is 'stationary'

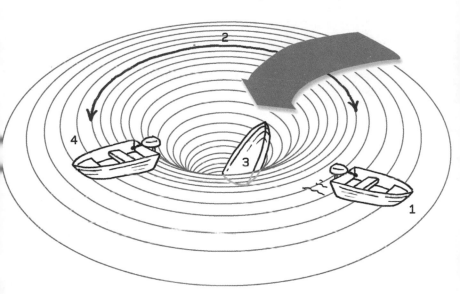

Neutron stars and pulsars

Neutron stars are extremely dense collapsed stars sometimes left behind after a core-collapse supernova explosion (see page 336). If the collapsing core is 1.4 to 3 times as massive as the Sun, it will form a neutron star, while a heavier core will collapse into a black hole (see page 346).

Neutron stars form because they are massive enough for their own gravity to compress normal matter into a superdense soup of neutrons surrounded by a solid crust of iron nuclei. Typically, they are about 15 km (9 miles) wide and spin very rapidly, sometimes once every few milliseconds. A teaspoon of material from the core of a neutron star would have a mass of roughly a billion tonnes. Neutron stars also have very intense magnetic fields, which accelerate particles into narrow polar beams emitting bright radiation.

Neutron stars called pulsars are easiest to detect because of their orientation. They happen to be aligned so that they sweep their bright radiation beams across Earth as they spin, so telescopes pick up regular pulses emerging from them.

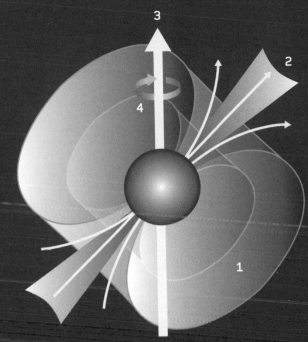

Structure of a pulsar

1 Powerful magnetic field around
  neutron star

2 Field channels radiation from star
  into narrow beams

3 Axis of rotation

4 Rapid spin of pulsar causes
  radiation beam to sweep
  across sky

Wormholes

A wormhole is a strange, hypothetical tunnel through space–time that could allow someone to take a shortcut from one place to another, apparently faster than the speed of light. No one has found any observational evidence that wormholes really exist, but Einstein's general relativity theory (see page 18) leaves open the possibility that they might.

A wormhole would involve a black hole (see page 346) connected to a hypothetical 'white hole' – an object that would act like the reverse of a black hole, allowing matter to come out, but nothing to get in. Jump into the black hole and you'd pop out again somewhere else in the universe, or even in another universe altogether.

To picture a traversable wormhole, think of a piece of paper bent in half without the two halves touching. A wormhole would be like a tunnel that connected the two sides with a shorter path than that following the curve of the paper, representing 'normal space'. Whether wormholes could really exist in nature is extremely doubtful, however.

Anatomy of a wormhole

1 Nearby region of space-time
2 Path of travel through normal space
3 Black hole
4 Wormhole
5 Hypothetical 'white hole'
6 Distant region of space-time

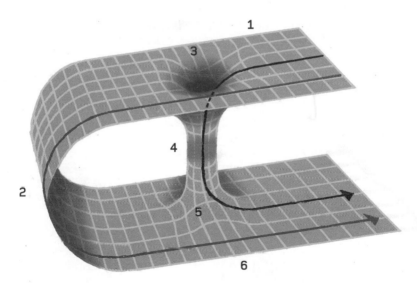

The Big Bang

The Big Bang was a cataclysmic explosion that created our universe around 13.7 billion years ago. The theory's credibility grew from the 1920s, when astronomers discovered that galaxies are receding from each other on large scales because the cosmos is expanding. That suggests all matter was much closer together in the distant past and points to the origin of the universe in a state of unimaginably high density.

Modern theories suggest that a split second after the Big Bang, a fleeting phase called cosmic inflation made the universe expand exponentially fast. After that, the dense fireball gradually cooled as it expanded more slowly, forming familiar particles such as protons and neutrons, then building atomic nuclei and finally neutral atoms within about 400,000 years. Regions with the highest density eventually collapsed under gravity to form galaxies of stars.

Much of our information about the early universe comes from the microwave background (see page 354). But it's still not clear what triggered the Big Bang in the first place.

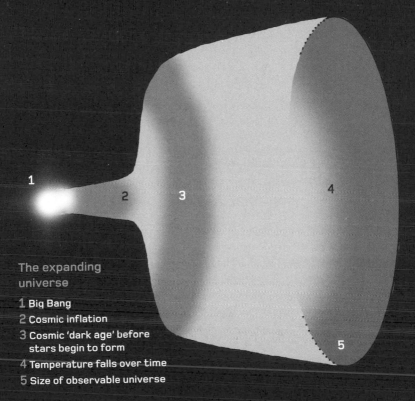

The expanding universe

1 Big Bang
2 Cosmic inflation
3 Cosmic 'dark age' before stars begin to form
4 Temperature falls over time
5 Size of observable universe

Cosmic microwave background

The cosmic background radiation is the afterglow of the Big Bang (see page 352). It pervades all space today and has been a vital tool in determining conditions in the early cosmos.

The Big Bang created an expanding fireball of enormous density that effectively trapped photons of light inside it. When the universe was 400,000 years old, however, the fireball had cooled enough for neutral atoms to form. Suddenly, the orange-red glow of heat from the fireball, now at about 3,000°C (5,400°F), could stream freely through the universe in every direction. We still see this radiation today, stretched into invisible microwaves by the universe's expansion.

The microwave background comes from every direction in the sky, a bit like cosmic wallpaper pasted behind all the galaxies. Satellite measurements reveal that it has subtle 'ripples' – tiny variations in wavelength – that arose due to the lumpiness of matter in the early universe. They encode amazingly rich information about the universe's history, including its age, expansion rate and composition.

A map of the entire sky from NASA's Wilson Microwave Anisotropy Probe (WMAP) reveals minute variations in the temperature of the microwave background radiation – dark areas indicate less dense regions of material in the early universe, while brighter ones mark dense patches that formed the seeds of the first galaxies

Universe

The universe is the totality of all space, matter and energy that exists. It formed in the Big Bang (see page 352) and since then galaxies have evolved within it, forming vast filaments that connect up in a giant cosmic web.

Most of the universe's mass/energy (73 per cent) is inexplicable dark energy (see page 362), while 23 per cent is unidentified dark matter (see page 360). Only about 4 per cent is normal matter, like that found in stars, planets and people.

Observations suggest the universe is at least 150 billion light-years wide. If the universe is finite, scientists favour the idea that it doesn't have edges. Instead, space would loop back on itself so that a rocket travelling in a straight line might eventually end up back where it started. Some models suggest the universe might take on one of many endlessly repeating shapes, including one based on a 12-sided dodecahedron.

Alternatively, the universe may be infinite, in which case it has always been infinite and the Big Bang took place throughout an infinite space.

A finite universe might seem like a hall of mirrors—a rocket travelling in a straight line might encounter the same space over and over again. For instance a rocket exiting a 'dodecahedron' universe on one face (1) might re-enter on its opposite face (2).

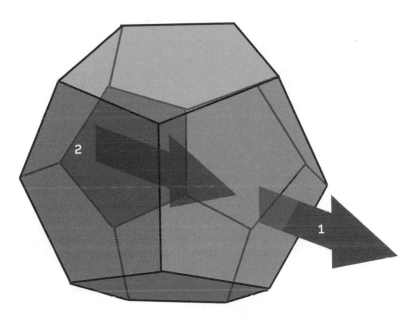

Gravitational lensing

Gravitational lensing occurs when the gravity of a foreground object bends and magnifies light from an object behind it, an effect that was predicted by general relativity (see page 18).

Dramatic gravitational lensing occurs when the huge gravity of a galaxy cluster magnifies the light of a galaxy behind it. Astronomers can use the effect as a 'zoom lens' to detect galaxies so distant that their light has taken more than 13 billion years to reach the Earth. Occasionally, a galaxy is so well lined up behind a cluster that the lensing effect distorts it into a neat circle called an Einstein ring.

On a smaller scale, a similar effect called microlensing can reveal new extrasolar planets (see page 338). When one star passes in front of another, the front star's precise distortion of the background one can carry subtle clues that a planet is orbiting the front star. Bizarrely, this allows astronomers to detect invisible planets circling invisible stars – the technique can work even if the front 'lensing' star is too faint to see.

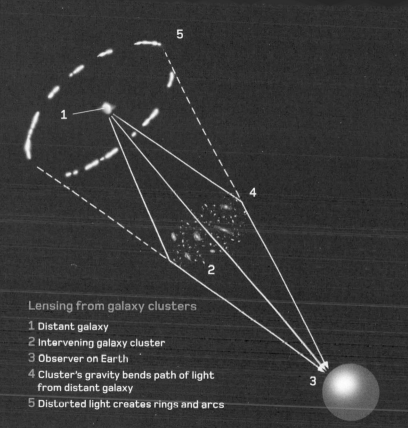

Lensing from galaxy clusters

1 Distant galaxy
2 Intervening galaxy cluster
3 Observer on Earth
4 Cluster's gravity bends path of light
   from distant galaxy
5 Distorted light creates rings and arcs

Dark matter

Dark matter is a mysterious invisible substance that makes up around 85 per cent of all the matter in the universe. Scientists know it's there only because it exerts a powerful gravitational force on visible stars and galaxies, influencing the way they move.

Since the 1930s, evidence has grown that stars in many galaxies move so fast that the galaxies should fly apart, unless they're held together by the gravity of dark, invisible matter. No telescope can see this dark matter. Unlike the normal atoms in stars, planets or people, it must be profoundly invisible and incapable of emitting or reflecting light. Dark matter may consist of 'weakly interacting massive particles', or WIMPS, that congregate into vast balls in and around galaxies.

An alternative explanation for this puzzle is 'modified Newtonian dynamics' (MOND), which assumes gravity's strength changes on large scales so dark matter isn't needed to explain star and galaxy motions. But no one-size-fits-all MOND theory so far explains all astronomical observations simultaneously.

1   2

Galaxies of bright stars **(1)** sit inside
vast balls of invisible dark matter **(2)**
that scientists have not yet identified.

Dark energy

Dark energy is a strange, unexplained effect that is causing the expansion of the universe to accelerate. Measurements suggest it's the dominant ingredient of the universe, accounting for 73 per cent of its total energy density.

The universe has expanded since the Big Bang (see page 352), and until the mid-1990s, astronomers assumed this expansion was gradually slowing down due to the attractive gravitational pull of all the matter inside, which resists expansion. But since then, studies of distant Type Ia supernovae (see page 336) have shown that they are dimmer than expected because the expansion of the universe has accelerated over time.

In other words, the universe contains 'dark energy' that is pushing galaxies apart. It might arise from a 'cosmological constant', a vacuum property that makes space 'springy'. Alternatively, space might be filled with an exotic 'quintessence', which acts as if it has a negative gravitational mass and hence causes repulsion. NASA and the European Space Agency are planning spacecraft missions to study dark energy further.

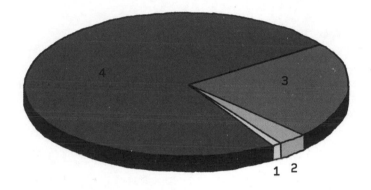

Ingredients of the universe

1 Ordinary atoms in galaxies: 0.4 per cent

2 Ordinary atoms in intergalactic gas: 3.6 per cent

3 Dark matter: 23 per cent

4 Dark energy: 73 per cent

Rocketry

Rocketry is the technology that enabled all the feats of the space age, including satellite launches, planetary probes and astronauts landing on the Moon. All rockets work by the principle of action and reaction in Newton's third law of motion (see page 10) – they push forwards by ejecting propellant backwards extremely fast. Most rockets burn liquid or solid fuel to achieve this.

The Second World War and the Cold War were driving forces for military rocket development and the subsequent space race. The German V-2 rocket, developed as a ballistic missile, is often considered the first object to have reached space on a suborbital flight, while a Soviet rocket launched the first satellite, Sputnik 1, in 1957. Human spaceflight began in 1961.

A rocket has to attain a specific speed, the so-called escape velocity, to overcome Earth's gravity and travel beyond our planet. From Earth's surface, the necessary escape velocity is about 11.2 km/s (25,000 mph) – roughly ten times faster than the record speed for a jet aircraft.

Escape velocity allows a projectile to overcome Earth's gravity

1 Not enough power to escape gravity

2 Enough power to go into orbit

3 Enough power to escape gravity altogether

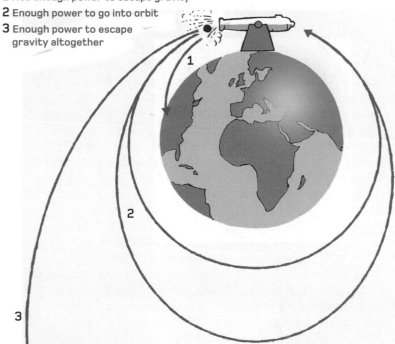

Satellites

Artificial satellites are spacecraft in Earth orbit (or probes orbiting another planet or the Moon). Today, there are more than 900 operational Earth satellites for purposes such as communications, navigation or weather forecasting.

Satellites are placed into orbit with a fixed speed – not so fast that they escape the Earth's gravitational pull, and not so slow that they fall back down to Earth. Many satellites, such as military reconnaissance satellites, are placed in low Earth orbits to get a close-up view of the Earth's surface.

Most communications satellites orbit in the 'geostationary ring' about 35,786 km (22,236 miles) above the Earth's equator. At this altitude, an orbit takes 24 hours, so a satellite hovers over a fixed point on the ground as the Earth rotates, maintaining a fixed communications link.

More than 5,000 tonnes of technology circles above our heads today. But most of it is useless 'space junk', such as discarded rocket stages, which threatens to damage operational satellites through collisions.

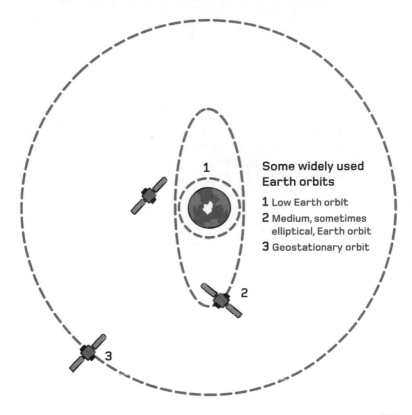

Some widely used Earth orbits

1 Low Earth orbit

2 Medium, sometimes elliptical, Earth orbit

3 Geostationary orbit

Planetary probes

Interplanetary spacecraft missions began in earnest in 1959, when the Soviet Union successfully crashed a probe, Luna 2, into the Moon. The Soviet Venera 7 probe was the first to beam back data from another planet after landing on Venus in 1970, while NASA's Mariner 9 spacecraft became the first to orbit another planet, arriving at Mars in 1971.

Several missions have robotically gathered extraterrestrial samples and returned them to Earth for analysis. Between 1970 and 1976, three Soviet missions returned samples of lunar soil. Other sample-return missions include NASA's Stardust project, which returned dust samples from a comet in 2006, and the Japanese Hayabusa mission, which returned asteroid samples in 2010.

Most of the Mars missions attempted in the late 20th century flopped, but success rates in the past decade have dramatically improved. NASA's robotic rovers Spirit and Opportunity operated on Mars for more than six years – more than 20 times longer than expected.

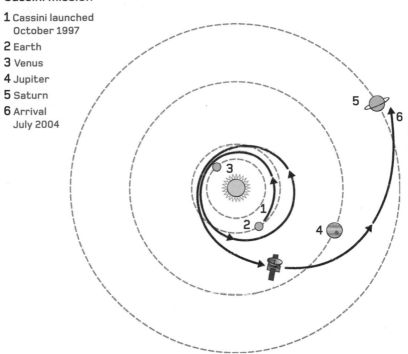

Flight path of NASA's Cassini mission

1 Cassini launched
   October 1997
2 Earth
3 Venus
4 Jupiter
5 Saturn
6 Arrival
   July 2004

Human spaceflight

Human spaceflight began in April 1961 with the flight of Yuri Gagarin on the Soviet Union's Vostok 1 spacecraft, which orbited the Earth once in 108 minutes. Gagarin's safe return laid to rest worries that spaceflight might be fatal for humans.

Alan Shepard became the first American in space the following month. NASA's extraordinary Apollo programme followed, landing the first men on the Moon in 1969. In total, 12 men walked on the Moon between 1969 and 1972.

The Soviet Union (later the Russian Federation) developed a strong track record for orbiting space stations, operating the Mir space station from 1986 until 2001. Astronauts often worked on Mir for nearly a year or more at a time.

NASA's Space Shuttle has flown crews into space more than 130 times. Five reusable shuttles originally existed, but accidents destroyed two in 1986 and 2003, claiming the lives of 14 astronauts. China became the third nation to independently launch astronauts into space in 2003, and many nations are collaborating on the current International Space Station.

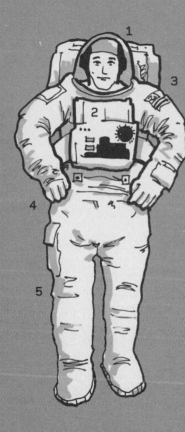

Components of a spacesuit for spacewalking

1 Helmet with radio for communication

2 Mini workstation with tools

3 Life support system, including oxygen supply

4 Gloves

5 Liquid cooling and ventilation garment with underlying pressure suit

Analogue and digital computing

Analogue computers are old-fashioned ones that work with continuously variable quantities like the strength of an electric current or mechanical rotation of a dial. Modern computers are based on digital technology. Information is represented as bits and bytes, sequences of binary 1s and 0s. Fundamental to the technology is the idea of on/off, true/false.

Analogue computers date back to ancient times, the oldest known example being the Greek Antikythera mechanism, which dates to between 150 and 100 BC and was designed to calculate astronomical positions. In the mid-1900s, scientists developed analogue computers with electrical circuits that could perform calculations. These computers were still in use in the 1960s and performed many of the calculations needed to plan NASA's Apollo spacecraft missions to the Moon.

Early digital computers used bulky 'thermionic valves' and later transistors to switch currents to perform calculations. Microchips (see page 374) revolutionized computer technology, paving the way for small, powerful desktop computers.

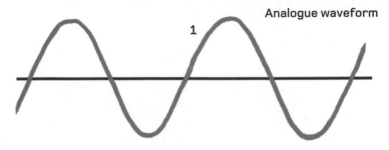

Analogue waveform

1

1 Continuously varying values
2 Information is quantized into steps

3 Binary data has only two values – 0 or 1

Digitized data

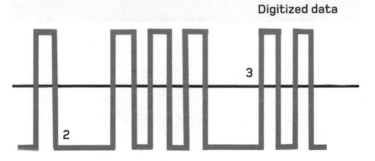

3

2

Microchips

A microchip or integrated circuit is a tiny electronic circuit on a semiconductor wafer. Jack Kilby at Texas Instruments in the US invented the first microchip in 1958, and today they are used in all common electronic devices including computers, mobile phones and sat-nav systems.

Digital computers perform calculations using transistors that can switch between two states, on and off, representing the binary digits 0 and 1. Microchips miniaturize the electronic circuitry required. They are cheap to make because the circuitry is 'printed' onto semiconductor wafers (see page 128) by photolithography, rather than being constructed one transistor at a time. A 'photo-resist' coating is applied to the wafer and ultraviolet light etches out the circuit pattern. Then another etching process lays down conducting metal paths.

Modern integrated circuits that are just 5 mm (0.2 in) square host millions of transistors, each much tinier than the width of a human hair. They can switch on and off billions of times a second.

Structure of a microchip

A complete microchip 'package' allows
the complex electronics of the integrated
circuit to interface with other elements
of a device:

1 Pins slot into printed circuit
board (PCB)

2 Fine 'bond wires' of aluminium,
copper or gold

3 Integrated circuit

4 Insulating substrate
surface

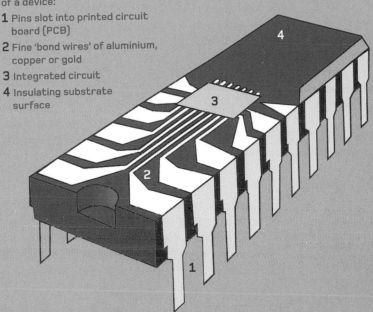

Computer algorithms

A computer algorithm is a sequence of instructions designed to solve a problem. It might specify the way a computer should calculate monthly payments for employees, for example, and how it should display the results. Real computer algorithms are normally very complicated, but this simple example outlines the steps to turn a daylight-sensing streetlamp on at night:

(1)   Is it dark? If yes, go to (2), if no, go to (3)
(2)   Turn on the light. Go to (3)
(3)   End

'Genetic algorithms' are ones that evolve in a process that mimics natural selection (see page 192). An algorithm designed to perform a certain task is tested and rated for its success, then allowed to 'breed' with other algorithms by mixing up their attributes. The most successful 'offspring' algorithms then breed and the process repeats until the computer 'evolves' the best algorithm for the job.

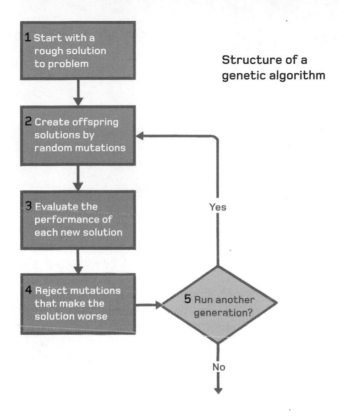

1 Start with a rough solution to problem

2 Create offspring solutions by random mutations

3 Evaluate the performance of each new solution

4 Reject mutations that make the solution worse

5 Run another generation?

Structure of a genetic algorithm

Yes

No

Neural networks

In computer science, a neural network is an information-processing concept inspired by the way biological nervous systems process information. Many processing elements are connected together like a network of biological neurons, and they work in unison to solve specific problems.

Conventional computers use algorithms (see page 376) to solve a problem, but this restricts their capabilities to problems we already know how to solve. Neural networks are like 'experts' in the information they analyse and are good at finding patterns in a large jumble of data. A neural network could compare the features of thousands of Hollywood movies in a database with their box-office takings, for instance, and pinpoint the factors that distinguish the hits from the flops.

Another application of neural networks is in face-recognition software. Computers can be trained to recognize a face by analysing images and comparing positions of features such as eye corners, but neural networks can *learn* which features are most useful for matching a face to images in a database.

A simple neural network consists of numerous interconnected processing units or 'synapses', each of which stores parameters known as weights. Data fed to input synapses is passed to one or more 'hidden' layers, whose weights are calculated using learning algorithms. The results of calculations by the hidden layers are then synthesized by output synapses.

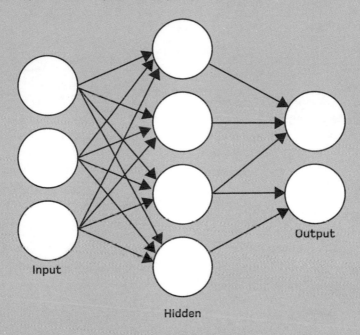

Input

Hidden

Output

Quantum computing

A quantum computer is one that would use the physics of quantum mechanics (see page 66) to increase its computational power beyond that of a conventional computer. Such instruments are still in a very early research phase.

Conventional computers store data in a binary series of 0s and 1s. Instead, a quantum computer would store information as 0, 1 or a quantum superposition of the two. These 'quantum bits', or qubits, would allow much faster calculations. While three conventional bits could represent any number from 0 to 7 at one time, three qubits could represent all these numbers simultaneously. This means a quantum computer could tackle many calculations simultaneously and solve problems that would keep today's supercomputers busy for millions of years.

Experimental quantum computers have used a few qubits to perform simple calculations like multiplying 5 by 3. It's not clear whether they will become a practical option because they rely on complicated and delicate procedures such as quantum entanglement (see page 74) to couple the qubits together.

One way physicists model the exact mixing of 1 and 0 in a quantum 'qubit' is to think of it as the latitude on a sphere, with a 'north pole' (1) equivalent to a value of 1 and a 'south pole' (2) equivalent to value 0. Superpositions – a mix of 1 and 0 values – can be considered as intermediate latitudes (3). The measurement process collapses the qubit into a classical value of 1 or 0, with the probability of each value given by the surface area on the opposite side of qubit's latitude (4).

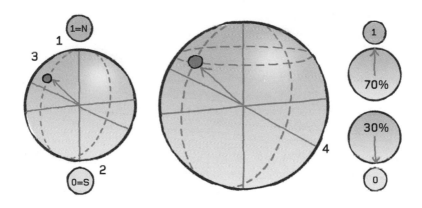

Turing test

The Turing test is a measure of a machine's ability to demonstrate intelligence. English mathematician, computer pioneer and Second World War code breaker Alan Turing proposed the test in the 1950s. Basically, it suggests that a computer has achieved human intelligence if it convincingly responds like a person.

Turing proposed an experiment in which a volunteer sits with an experiment manager behind a screen. On the other side, out of sight, a second volunteer asks questions. The first volunteer and a computer both answer with text messages, and the manager decides at random which of the two responses the questioner will receive. If the questioner can't distinguish the human responses from the computer ones, the computer has achieved human intelligence.

Turing predicted that machines would eventually pass the test. Various commercial text and email programs regularly trick people into thinking they've communicated with a person, but no computer has yet passed a rigorous Turing test.

Experimental set-up for a Turing test

1 Human questioner
2 Terminals to display responses
3 Barrier
4 Human answerer
5 Computer
6 Experiment controller relays either human or computer response

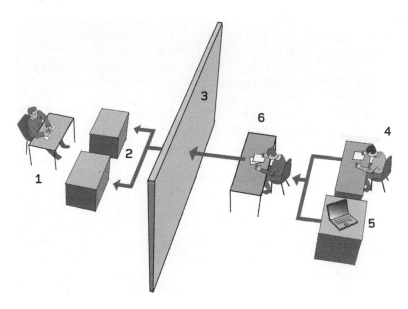

Hard drives

Hard disks in computers and servers store changing digital information in a fairly permanent form, giving computers the ability to 'remember' data even when they're switched off. They consist of several solid disks or 'platters' on which data are stored magnetically, and a read/write head to record and retrieve information.

The technology was invented in the 1950s and later took the name 'hard disks' to distinguish them from floppy disks, which stored data on flexible plastic film. The platters in a hard disk are usually made of aluminium or glass with a coating of magnetic recording material, which can be easily erased and rewritten and preserves information for many years.

When the drive is operating, the platters typically spin 7,200 times a minute. The arm holding the read/write heads can often move between the central hub and the edge of the disk and back up to 50 times per second. Some desktop computers now have hard disks with more than 1.5 terabytes (1.5 million million bytes) of memory.

Structure of a hard drive

1 Spindle
2 Stacked platters
3 Movable read/write head
4 Controller electronics
5 Casing and interface

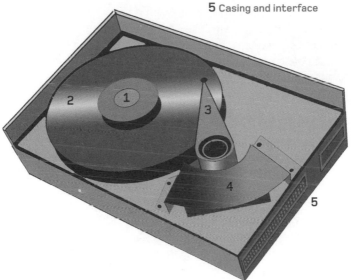

Flash memory

Like hard disks (see page 384), flash memory stores digital information and 'remembers' it even when the power is switched off. Unlike hard disks, flash memory does not have any moving parts. It doesn't mind a good hard knock and can withstand large temperature variations, and sometimes even immersion in water. That makes it the ideal memory for portable devices.

Flash memory works by switching transistors on and off to represent sequences of 0s and 1s. Unlike conventional transistors, which 'forget' information when the power is off, the transistors in flash memory have an extra 'gate' that can trap electric charge, registering a 1, until another electric field is applied to drain the charge and return the bit to 0.

Flash memory is used in mobile phones, MP3 players, digital cameras and memory sticks, which are often used to back up files or transfer files between computers. Some memory sticks have storage capacities of 32 gigabytes, enough to store around 20 hours of video.

Structure of a flash memory 'cell'

1 Source line for electric current

2 Insulated float gate stores information as electric charge

3 Control gate governs flow of electricity from source to drain, determined by charge on float gate

4 Drain line for current

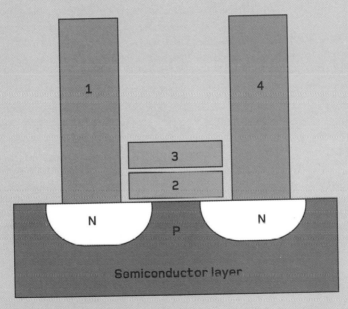

Optical storage

Optical storage refers to types of memory, such as CDs and DVDs, that are read by a laser. Today, desktop computers have drives that both read and write these media.

Both CDs and DVDs have a long, spiralled track up to around 12 km (7.5 miles) long. Mass-produced CDs and DVDs have little bumps around the track that encode digital data as a series of 0s and 1s. To read the data, a red laser bounces light off the bumps and a sensor detects height changes by measuring the reflected light.

CD burners are now standard in personal computers. Write-once CDs are coated with a layer of see-through dye, and a laser burns data onto the disc by turning this dye opaque. Rewritable CDs use a more complicated chemical trick that allows data to be erased again by laser heating. Blu-ray discs can store even more information than DVDs because they are read with a blue-violet laser that has a shorter wavelength than a red laser, making it possible to focus the laser spot with much greater precision.

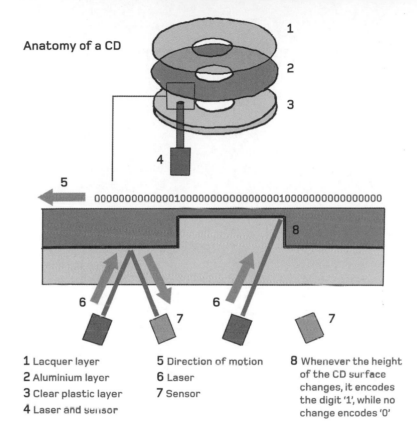

Anatomy of a CD

000000000000010000000000000000010000000000000000

1 Lacquer layer
2 Aluminium layer
3 Clear plastic layer
4 Laser and sensor

5 Direction of motion
6 Laser
7 Sensor

8 Whenever the height of the CD surface changes, it encodes the digit '1', while no change encodes '0'

Holographic memory

Holographic memory might one day revolutionize high-capacity data storage. Today, magnetic storage and optical storage (see page 388) are the usual ways of storing large amounts of data, recording individual 'bits' on a surface and reading them one bit at a time. The holographic technique would record information in a 3D volume and read out millions of bits simultaneously, speeding up data transfer enormously.

To record holographic data, a laser beam is split into two, with one ray passing through a filter carrying raw binary data as transparent and dark boxes. The other 'reference' beam takes a separate path, recombining with the data beam to create an interference pattern (see page 64), recorded as a hologram inside a light-sensitive crystal. To retrieve the data, the reference beam shines into the crystal at the same angle that it stored the data, to hit the correct data location inside.

Several companies hope to develop commercial holographic memory, which could one day store many terabytes (millions of millions of bytes) of data in a crystal the size of a sugar cube.

Holographic data storage

1 Laser source generates beam
2 Splitter produces two identical laser beams from the original one
3 Filter modifies beam, encoding data to be stored
4 Mirror redirects unmodified beam into recording block
5 Data storage device records interference between modified and unmodified beams

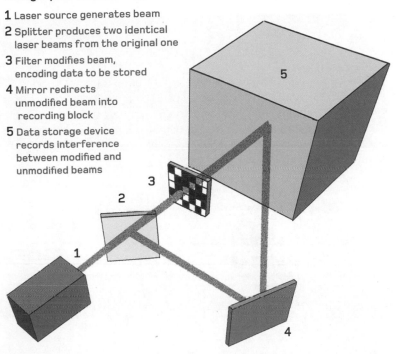

Radar

Radar is a technique for detecting objects and measuring their distances and speeds by bouncing radio waves off them. It developed rapidly during the Second World War and is still used in a wide range of applications, including air traffic control and weather forecasting as well as satellite mapping of the Earth's terrain and that of other planets.

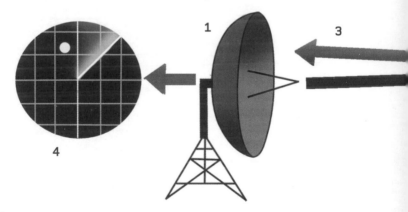

Radar stands for 'radio detection and ranging'. A radar dish, or antenna, transmits pulses of radio waves or microwaves that reflect off any object in their path. The reflected part of the wave's energy returns to a receiver antenna, with the arrival time indicating the object's distance. If the object is moving towards or away from the radar station, there is a slight difference in the frequencies of the transmitted and reflected waves due to the Doppler effect (see page 42).

Marine radars on ships prevent collisions with other ships, while meteorologists use radar to monitor precipitation. Similar systems operating with visible laser light are called lidar, and can measure details with higher precision.

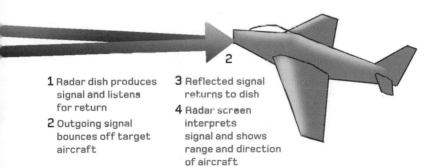

1 Radar dish produces signal and listens for return

2 Outgoing signal bounces off target aircraft

3 Reflected signal returns to dish

4 Radar screen interprets signal and shows range and direction of aircraft

Sonar

Sonar is a technique that ships use to navigate and detect other vessels, or to map the ocean floor, using sound waves. 'Passive' sonar instruments listen out for the sounds made by other ships or submarines, while 'active' sonar systems emit sound waves and listen for the echoes.

Sonar stands for 'sound navigation and ranging', and the first instruments developed rapidly during the First World War for detection of enemy submarines. Active sonar creates a pulse of sound, often called a ping, and then listens for reflections of the pulse, the arrival time of the reflections indicating the distance of an obstacle. Outgoing pings are single-frequency tones or changing-frequency chirps, which allow more information to be extracted from the echo. Differences in frequency between pings and echoes can allow measurement of a target's speed, thanks to the Doppler effect (see page 42).

Fishing boats use sonar to pinpoint fish shoals, while some animals, including bats and dolphins, use similar natural echo-location to navigate or locate mates, predators and prey.

1 Sonar pod on fishing trawler generates sound waves

2 Shoal of fish

3 Time for sound waves to return to ship is measured

Internet and World Wide Web

The Internet is a global system of interconnected computers that use the 'Internet Protocol Suite' as a common language to speak to each other. It's a vast network formed by myriad smaller networks run by organizations including private companies, universities and government bodies, linked together by fibre-optic cables, phone lines and wireless technologies.

The World Wide Web, or usually just the Web, is a way of handling documents over the Internet. Web browser software allows users to view pages containing text, images, videos and other multimedia and jump between them via 'hyperlinks'. English computer scientist Tim Berners-Lee is credited with inventing the Web in 1989 while at CERN, the European centre for particle physics on the French–Swiss border.

The main mark-up language for Web pages is HTML (hypertext mark-up language), which uses tags at either end of text phrases to tell a Web browser how to display them – for instance, as a clickable hyperlink. Estimates suggest more than 2 billion people worldwide currently access the Web.

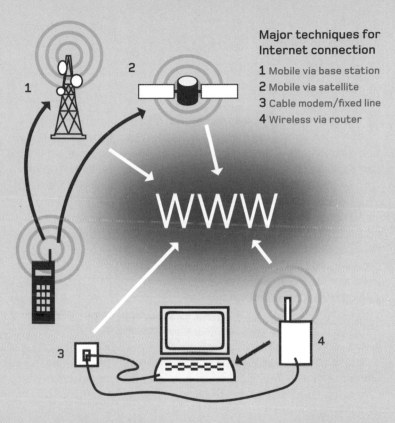

Major techniques for Internet connection

1 Mobile via base station
2 Mobile via satellite
3 Cable modem/fixed line
4 Wireless via router

WWW

Internet security

The Internet allows easy transfer of information, but it also allows the spread of 'malware' – programs written with malicious intent. Computer viruses are harmful programs that can transfer between computers via email, aiming to delete files or disable operating systems like Microsoft Windows.

Other malware includes 'spyware', which might stealthily install itself on a computer and transmit the user's secret passwords to fraudsters, while a computer 'worm' self-replicates and sends copies of itself to other computers on a network. Networked computers need constantly updated anti-virus software to detect and remove new malware, as well as 'firewalls' that prevent unauthorized access from the outside.

Denial-of-service attacks attempt to make an organization's website useless by bombarding it with so many communication requests that it can't cope with legitimate traffic. People install software agents called bots to launch attacks on specific sites, or make bots infect computers by stealth. Many countries deem denial-of-service attacks a criminal offence.

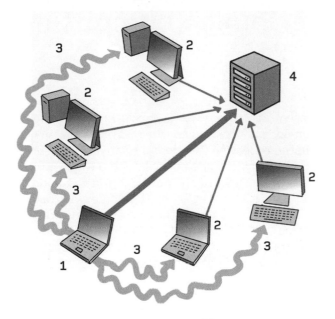

In a denial-of-service attack, the attacker (1) hijacks the computers of other users (2) by spreading malware (3). On command, the resulting 'botnet' bombards a remote server computer (4) with requests for information, overloading it.

Distributed computing

A distributed computing project is one that uses many different computers working together to solve a problem, with each computer taking charge of a small piece of the overall data processing. The goal is to complete the task much faster than would be possible with a single computer.

One type of distributed computing is grid computing, in which many computers cooperate remotely, sometimes using the idle time of ordinary home computers. An example of this is the 'SETI@Home' project launched in 1999. Around 8 million people have signed up to download a screensaver-like program that sifts little packets of data from the Arecibo radio telescope in Puerto Rico to look for unusual signals – some of which might be communications from intelligent alien civilizations – and return the results to project organizers.

Folding@home is a similar computing project that invites the public to use their computers to analyse protein folding (see page 138). This could provide vital information that leads to new treatments for diseases such as cancer and Alzheimer's.

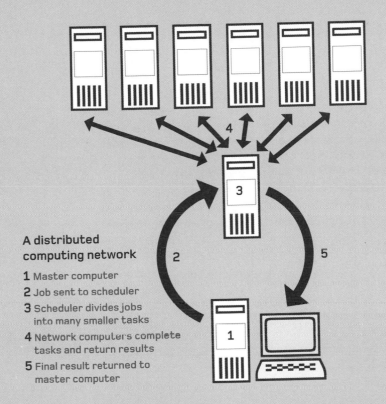

A distributed computing network

1 Master computer
2 Job sent to scheduler
3 Scheduler divides jobs into many smaller tasks
4 Network computers complete tasks and return results
5 Final result returned to master computer

Speech communications

The 1870s saw the invention of the telephone, when Scottish-born US scientist Alexander Graham Bell transmitted speech electrically down a wire. A microphone in a handset vibrated in response to sound, creating an electrical signal by induction (see page 50) that travelled down a wire to cause the reverse process in a loudspeaker, with the current making the speaker vibrate to reproduce the sound.

The first commercial mobile phones were launched in the late 1970s, relaying signals wirelessly to local transmitters that pass them on to the main landline network. Most phone signals today are digital, encoded as a series of 0s and 1s. The past decade has seen rapid growth of phone calls over the Internet (Voice over Internet Protocol or VoIP), which has slashed the costs of long-distance calls.

Satellite phones are used in remote regions where there is no mobile phone signal or landline network. They communicate directly with an overhead satellite, which beams the signal back to a ground antenna where a public phone network is available.

How a satellite phone works

1 Phone unit
2 Orbiting communications satellite
3 Gateway groundstation
4 Public telephone network
5 Landline telephone

Fibre optics

Optical fibres are thin, flexible strands of transparent material used to 'pipe' light signals around a network, transmitting all kinds of data including Internet traffic and phone calls. These fibres can allow faster data transfer than conventional electrical cables and can transmit signals for tens of kilometres without any amplification.

A single fibre has a thin glass or plastic core with an outer cladding of optical material that constantly reflects light back into the fibre to confine it, a process called total internal reflection. An outer plastic coating protects the fibre from moisture and damage. Typically, hundreds or thousands of fibres are bundled together in cables with an outer jacket.

Single-mode cables transmit one wavelength of light through cores thinner than a human hair, while multimode cables have wider cores that can transmit several different wavelengths. Light signals travel through them at around 200,000 km/s (450 million mph), allowing phone conversations to anywhere in the world without an annoying delay or echo on the line.

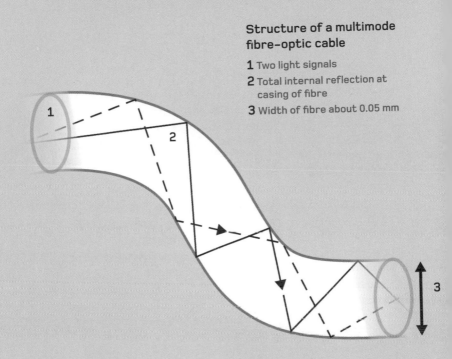

Structure of a multimode fibre-optic cable

1 Two light signals

2 Total internal reflection at casing of fibre

3 Width of fibre about 0.05 mm

GPS

The Global Positioning System, or GPS, is a network of satellites maintained by the US government that tells receivers on the ground their precise location. Anyone with a GPS receiver, or sat-nav, can freely access it. Russia also has a satellite navigation system called GLONASS, while China and the European Union both have plans for new ones.

A sat-nav receiver calculates its position by timing signals sent by four or more GPS satellites, which tell the receiver when and where the signals were emitted. This position is shown on a screen, often with a moving map display.

At any one time there are more than 24 active GPS satellites operating in medium Earth orbit. As well as road vehicles, users include map makers, aircraft and ships. GPS also allows electronic monitoring of criminals under curfew or even pets, using devices that locate themselves using GPS and report their position via a mobile phone network, for instance. Some GPS communications are encoded for military use only.

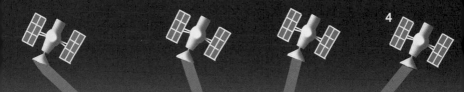

GPS at work

1 GPS satellite receiver on Earth receives four signals simultaneously, each of which was sent at a slightly different time

2 Signals from closest satellites reach receiver in shortest time

3 Signal from more distant satellite takes longer to reach receiver

4 Signals from three satellites reveal GPS user's location on Earth's surface—four signals reveal altitude as well

Glossary

Algae
Simple plants that lack specialized parts such as roots; includes single cells and large aquatic plants such as kelp.

Antenna
Device that transmits/receives electromagnetic waves by converting waves into electric current or vice versa.

Bell curve
A bell-shaped distribution in which most values cluster near the mean value but tail off above and below that.

Binding energy
The energy required to separate all the constituent particles in a molecule, atom or atomic nucleus.

Buoyancy
An upward force, caused by pressure of a liquid or gas, that tends to make objects float.

Cell
In biology, the basic structural unit of living organisms from single-celled bacteria to plants and animals.

Condensation
The process of changing from a gas to a liquid or solid.

Conductor
A substance that readily conducts energy, such as electrical current, heat or sound waves.

Crest
In physics, the peak of a waveform, as opposed to the trough, the lowest part.

Cytoplasm
The jelly-like internal medium of a cell outside the nucleus.

Ecosystem
A community of organisms linked by their interactions with each other and their environment.

Eddy
In fluid dynamics, the swirling of a fluid and reverse current created when the fluid flows past an obstacle.

Electron
A stable subatomic particle with negative electric charge; one of the fundamental particles in nature.

Evaporation
Vaporization from a liquid's surface to form a gas, for instance as puddles dry out in warm sunshine.

Free radical
Highly reactive, unstable atoms or molecules with at least one unpaired electron that can damage cells.

Friction
The force that resists motion of objects or fluids moving past each other, often dissipating energy as heat.

Galaxy
A gravitationally bound system of millions or billions of stars and their associated gas and dust.

Germ cell
A reproductive cell such as sperm or eggs involved in transmitting DNA to an organism's offspring.

Glands
Organs that synthesize substances needed by the body, releasing them through ducts or into the bloodstream.

Greenhouse gas
Gases such as carbon dioxide that absorb solar radiation and trap heat in the atmosphere.

Humidity
A measure of the amount of water vapour in the atmosphere; often highest in tropical forests.

Inflammation
An immune response that can cause redness/swelling; can prevent infections or initiate healing of injuries.

Insulator
A substance that does not readily conduct energy such as electrical current, heat or sound waves.

Lattice
An arrangement of points, particles or objects in a regular periodic pattern in two or three dimensions.

Mass
A property of matter that gives it weight in a gravitational field; energy also has an associated mass.

Mineral
A naturally occurring solid formed through geological processes with a characteristic chemical composition.

Molecule
A group of atoms of one or more elements that are chemically bonded together in a definite arrangement.

Neuron
A nerve cell that is specialized to conduct nerve impulses by electrical and chemical signalling.

Neutron
An electrically neutral subatomic particle usually found in atomic nuclei; composed of three quarks.

Organic
In chemistry, a term describing the huge range of chemical compounds based on the element carbon.

Organism
Any living thing, such as an animal, plant, fungus or microbe, that consists of one or more cells.

Phytoplankton
A photosynthetic or plant constituent of plankton; usually refers to single-celled algae.

Plasma
A electrically conducting hot gas that's partially ionized; also the watery fluid that carries blood cells.

Pollen
Tiny grains made by male part of a flower; combines with female part in pollination for seed production.

Proton
A positively charged subatomic particle usually found in atomic nuclei; composed of three quarks.

Radiation

Travel of energy through a medium or through space; usually refers to electromagnetic waves.

Red giant

A giant star of low to medium mass (roughly 0.5 to 10 solar masses) in a late phase of stellar evolution.

Seismic waves

Vibrations that propagate through the Earth (or another body) following an earthquake or explosion.

Space–time

The inextricably interwoven dimensions (three of space and one of time) in Einstein's relativity theories.

Superposition

In quantum mechanics, the overlapping of waves that represent the various possible states of a particle.

Tetrahedron

Any polyhedron with four plane faces; a regular tetrahedron's faces are four equilateral triangles.

Toxin

A poison, usually a protein made by a pathogenic bacteria, that is highly toxic for other living organisms.

Trough

In physics, the bottom or lowest part of a waveform, as opposed to the crest, the highest part.

Turbine

A rotary engine in which a moving fluid rotates a bladed rotor to provide mechanical energy.

Turbulence

Disturbance in a fluid (liquid or gas) characterized by chaotic flow; can impede motion of vehicles.

Vacuum

A volume of space that's nearly empty of matter, so its gas pressure is much less than atmospheric pressure.

Vortice

A spinning, rapid and often turbulent flow of fluid around a centre with a pressure minimum at its centre.

Index

Quercus Editions Ltd
55 Baker Street
7th Floor, South Block
London
W1U 8EW

First published in 2011

A catalogue record of this book is available from the British Library

UK and associated territories
    Hardback edition: 978 0 85738 615 1
    Paperback edition: 978 1 78087 144 8
    Promotional paperback edition: 978 1 78206 251 6

Printed and bound in China

10 9 8 7 6 5 4 3 2